MILLIONS, BILLIONS, ZILLIONS

Also by Brian W. Kernighan:

The Elements of Programming Style (with P. J. Plauger)

Software Tools (with P. J. Plauger)

Software Tools in Pascal (with P. J. Plauger)

The C Programming Language (with Dennis Ritchie)

The AWK Programming Language
(with Al Aho and Peter Weinberger)

The Unix Programming Environment (with Rob Pike)

AMPL: A Modeling Language for Mathematical Programming
(with Robert Fourer and David Gay)

The Practice of Programming (with Rob Pike)

Hello, World: Opinion Columns from the Daily Princetonian

The Go Programming Language (with Alan Donovan)

Understanding the Digital World

MILLIONS
BILLIONS
ZILLIONS

DEFENDING YOURSELF IN A WORLD

OF TOO MANY NUMBERS

BRIAN W. KERNIGHAN

PRINCETON UNIVERSITY PRESS
PRINCETON AND OXFORD

Published by Princeton University Press
41 William Street, Princeton, New Jersey 08540

In the United Kingdom: Princeton University Press
6 Oxford Street, Woodstock, Oxfordshire OX20 1TR

press.princeton.edu

Library of Congress Control Number: 2018942714

ISBN 978-0-691-18277-3

British Library Cataloging-in-Publication Data is available

Editorial: Vickie Kearn and Lauren Bucca   Production Editorial: Nathan Carr
Jacket/Cover Design: Lorraine Doneker   Production: Erin Suydam
Jacket image courtesy of 123RF   Proofreader: Wendy Washburn
Publicity: Sara Henning-Stout and Kathryn Stevens

Camera-ready copy for this book was produced by the author
in Times Roman and Helvetica, using groff, ghostscript,
and other open source Unix tools.

Printed on acid-free paper. ∞

Printed in the United States of America

10 9 8 7 6 5 4 3 2 1

For Meg and Mark

Contents

Preface

"When you have mastered numbers, you will in fact no longer be reading numbers, any more than you read words when reading books. You will be reading meanings."

(W. E. B. Du Bois, sociologist, writer and civil rights activist)

"You don't have to be a mathematician to have a feel for numbers."

(John Nash, mathematician and Nobel Prize winner)

"On average, people should be more skeptical when they see numbers. They should be more willing to play around with the data themselves."

(Nate Silver, statistician)

We are surrounded by numbers. Computers produce them at a furious rate, and they are passed along by politicians, reporters and bloggers, and included in the endless advertising barrage that we are all subjected to. Indeed, the flow of numbers is so high that most people (including me) can't cope, and our brains tune it all out. At best we take away a vague impression that something is important and should be believed because it has numbers attached to it.

Tuning out is a bad strategy, however, since most of those numbers are intended to convince us of something, whether to behave a particular way, or believe some politician, or buy a gadget, or get something to eat, or make an investment.

The goal of this book is to help you to assess the numbers that you encounter every day and to be able to produce numbers of your own when you have to, for your own good or merely as a counterweight to what you are being told by others. You should be able to recognize potential problems in what you hear, and to be leery of taking too much at face value.

This book will help you to be intelligently skeptical about the numbers that you see, to be able to reason about them, to decide whether some claim might be true or is clearly false, and to compute your own numbers when you need them to make important decisions. The general approach is to look at an example of some numeric value that's clearly, or at least probably, wrong, show how you can deduce that it's wrong, help you come up with your own number that's more likely to be right, and finally to draw some general lessons.

Once you've been properly armed, there are many different ways in which you can look after yourself. The primary thing you need is common sense, augmented by healthy skepticism, basic facts and a few ways of reasoning. It will help if you become comfortable doing approximate arithmetic—few problems need precise computation—and there are shortcuts to make it easier; we'll talk about them along the way.

The book is aimed at anyone who wants to be better informed and a lot more cautious about believing what they are being told. There is so much misinformation and deliberate falsehood today that we really have to pay attention if we hope to spot the errors, the outright lies, and the subtle misrepresentations and exaggerations.

None of this is rocket science, and it's not "mathematics" either. I've heard all too many people say "I was never any good at math." They are being unfair to themselves. What that really means is that they were not well taught, and never got much chance to use simple arithmetic in the service of day-to-day life. The material here requires nothing beyond elementary-school arithmetic. If you got to grade 5 or 6 in the United States or the equivalent somewhere else, that's all the technical or math background you need. After that, you just have to use your head and what you already know. You may even find that it's fun.

Acknowledgments

I am deeply indebted to Jon Bentley for his detailed comments on literally every page of multiple drafts of the manuscript. The book is much the better for Jon's contributions.

Paul Kernighan provided a number of fine examples, and his eagle eye spotted an embarrassingly large number of typographical errors; any remaining typos are entirely my fault.

I am also grateful for helpful suggestions from Josh Bloch, Stu Feldman, Jonathan Frankle, Sungchang Ha, Gerard Holzmann, Vickie Kearn, Mark Kernighan, Harry Lewis, Steve Lohr, Madeleine Planeix-Crocker, Arnold Robbins, Jonah Sinowitz, Howard Trickey and Peter Weinberger. The production team at Princeton University Press—Lauren Bucca, Nathan Carr, Lorraine Doneker, Dimitri Karetnikov and Susannah Shoemaker—has been a pleasure to work with.

As always, my profound thanks to my wife Meg for insightful comments on the manuscript, and for her support, enthusiasm and good advice for many years.

I would also like to acknowledge the newspapers and magazines, especially the *New York Times*, that have provided many of the examples here. They occasionally make mistakes, but they publish corrections. In a time of far too much "fake news" and outright lies, it's invaluable to have sources that care so much about truth and accuracy.

The web site at millionsbillionszillions.com has examples that didn't make it into the book, and new ones will be added over time. Please send along anything you find; I'd love to hear from you.

MILLIONS, BILLIONS, ZILLIONS

Chapter 1

Getting Started

"How many @#$%^&* cars *are* there?"
(the author, stuck in yet another endless traffic jam)

I've asked myself this question any number of times when I'm in a traffic jam with no end in sight, just immobile cars as far as the eye can see. This has happened to me in the United States, Canada, England and France over the past few years; you've undoubtedly had similar experiences somewhere.

So how many cars are there? You might wonder about the number on the road ahead, or in the town you live in, or the total in your home country.

Stop right now! Don't reach for your computer or your phone; don't ask Siri or Alexa. Imagine that you're in a situation where you simply can't ask. Perhaps your traffic jam is out in the countryside where there is no cell service, or you're on a plane without Internet connectivity, or maybe you're in an interview where a prospective employer wants to see whether you can think for yourself.

Figure 1.1: How many cars *are* there?

Your job is to figure out sensible answers on your own, without consulting any other sources—in other words, to come up with an *estimate*. Dictionary.com defines the noun *estimate* as "an approximate judgment or calculation, as of the value, amount, time, size, or weight of something," and the verb as "to form an approximate judgment or opinion regarding the worth, amount, size, weight, etc., of." That's exactly what you should do first.

Make your own estimate first.

As a specific example, let's estimate the number of cars in the United States. The way we approach this would be the same in other parts of the world, though the details might well be different.

It's often easiest to work from the bottom up, starting from something concrete that you know or have experience with, then build on that to the general situation. I would start with my own experience: there are three people in my immediate family, and we each have one car. If it were that simple—one car per person—then we're done. The population of the US today is around 330 million, so there are 330 million cars. And for many purposes, that estimate would be entirely adequate.

A rough estimate is often good enough.

Notice that our estimate came from two things: personal experience and knowledge of a single fact, the approximate population of the country. As we'll see in the rest of the book, we can make remarkably good estimates without detailed knowledge, but in the end, we do have to know something.

The more you know, the better your estimates.

330 million is probably too high, since many people don't have cars—children up to some age like 18 or 20, elderly people who no longer drive, and of course those who live in big cities where parking is expensive and public transportation is good. On the other side, some people will have more than one car, but that's likely to be much less common.

Taking such factors into account, we can refine the estimate of 330 million. If more than half of the population in the US has a car, perhaps even two thirds or three quarters, that leads to a more refined estimate of 200 to 250 million cars.

Refine your estimate if necessary.

Don't forget the "if necessary" part. Very often, a rough answer is entirely adequate, and sometimes there's no way to get information that would make it possible to refine anyway. We'll see plenty of examples of that along the way, and Chapter 13 offers some advice and a chance to practice.

We'll also see examples where people claim more knowledge and more accuracy than they could possibly have; that's a sign of something sketchy going on. If you've done your own estimation before accepting someone else's values, you'll be alert for such situations.

If we now turn to a computer or phone, we can compare our estimate to other sources. For instance, Wikipedia says that "there were an estimated 263.6 million registered passenger vehicles in the United States in 2015." The top Google hit is a story from the *Los Angeles Times* that says 253 million cars. Our estimate is certainly close to those, which is an encouraging sign.

Independent estimates should be similar.

Consensus is a good sign, unless everyone is making the same error. If two independently created estimates are significantly different, however, something is awry and at least one of them is wrong.

Now that we have a decent value for the number of cars, we can think about related questions. For example, how many miles does a typical car get driven in a year? How long does it last? How many cars are sold each year? How much does it cost to own a car?

How many miles does a car travel in a year? As above, it's useful to start with personal experience or observation. For instance, suppose you or some family member commutes to work 20 miles each way. That's 200 miles in a week, and thus about 10,000 miles in a 50-week year. Again, there will be a lot of variability: some people will have longer commutes, while others will have shorter, and still others will use public transportation. Short weeks and vacation trips and who knows what else will change the estimate to some extent, but many of these effects will average out.

Too big and too small tend to average out.

My auto insurance policy says that the premium for each of the family cars is based on an average of 27 miles per day, which on the surface is a strange number. But 365 times 27 is 9,855, which is close to 10,000. I suspect that this is not a coincidence: the insurance company knows that 10,000 miles per year is a good representative value.

How long does a car last? I've owned quite a few cars over the years, tending to drive them until they really start to fall apart; my last one was 17 years old and had 180,000 miles. I probably hang on longer than the norm, so we might choose a round number, say 100,000 miles or 10 years, though this is definitely a rough estimate. What about people who lease a new car every few years? When they upgrade, someone else will acquire a lightly used car, and it will go on to a typical lifetime, so 10 years is still reasonable.

How many new cars are sold each year? If there are 250 million cars and each lasts 10 years, then one tenth of them, or about 25 million, must be replaced each year; if instead they last 15 years, then 16 or 17 million will be replaced.

This is an example of a kind of conservation law: a car that reaches the end of its life will generally be replaced by a new one. Of course that assumes a steady state, which isn't true with a growing population or fluctuating economic conditions, but it's a reasonable assumption for getting started. Chapter 7 talks more about conservation laws.

Conservation: what goes in must come out.

How much does it cost to own a car? For practice, you might estimate how much per mile it costs to drive. This includes variable costs like fuel, fixed costs like insurance, unpredictable costs like repairs, and the money you'll need for a new car when the old one dies.

You may have noticed that for all of the estimates above, we used no arithmetic operations more complicated than multiplication and division, and we rounded off numeric values ruthlessly to make the computations easy.

**Multiplication, division and
approximate arithmetic are good enough.**

This will be true of the rest of the book as well—we're not doing "mathematics," but elementary-school arithmetic in a really relaxed way. Chapter 12 has a more extended discussion of arithmetic, with some shortcuts and rules of thumb to make it easier.

I've written this chapter mostly in terms of cars, which may not be of direct interest to you. But even if not, in later chapters we'll see that the approaches and techniques are applicable to any situation where you have to estimate something from incomplete information. Most of the time, you can get a value by looking it up, but you will be much better off if, before you resort to a search engine, you make your own estimate. It won't take long and you'll quickly get good at it. Practice will arm you for a lifetime of being wary about what people are telling you. If you have already thought about some numeric value and have done a bit of easy arithmetic, it's much less likely that someone can put something over on you.

A note on units

Given the accident of where I live, the majority of the examples in this book come from American sources. I'm not too worried about this fact; any part of the world will have similar stories.

I'm more concerned that the units of measure in many examples—lengths, weights, capacities—are expressed in "English" units because the US is almost unique in not adopting the metric system, and still uses English units for pretty much all weights and measures. Readers who are not familiar with feet, pounds and gallons may sometimes be perplexed by unfamiliar units. I've tried to alleviate this when possible, but failing to get the units right is often the point of the story.

Meanwhile, here's a list of the most common English units that appear in the book, with approximate conversions to and from metric units.

1 inch = 2.54 cm	1 cm = 0.3937 inches
1 foot = 12 inches = 30.48 cm	1 meter = 3.28 feet
	= 39.37 inches
1 yard = 3 feet = 0.9144 meters	1 meter = 1.09 yards
1 mile = 5,280 feet = 1,609 meters	1 km = 0.62 miles
	= 3,281 feet
1 ounce = 28.3 grams	1 gram = 0.035 ounces
1 pound = 16 ounces = 453.6 grams	1 kg = 2.204 pounds
1 ton = 2,000 pounds = 907.2 kg	1 metric ton = 1,000 kg
	= 2,204 pounds
1 US pint = 16 fluid ounces = 0.47 liters	1 liter = 2.11 US pints
1 gallon = 4 quarts = 8 pints = 3.79 liters	1 liter = 0.26 gallons
	= 1.06 quarts

1 acre = 0.405 hectares	1 hectare = 2.47 acres
1 square mile = 640 acres	1 hectare = 0.0039 square miles
1 Fahrenheit degree = 5/9 Celsius degrees	1 Celsius degree = 1.8 Fahrenheit degrees

   If you look at these conversions carefully, you can see some handy approximations:

   1 meter ~ 1 yard
   1 kilogram ~ 2 pounds
   1 liter ~ 1 quart
   1 Celsius degree ~ 2 Fahrenheit degrees

These are all within 10 percent of the true values. If you really need more accuracy, then add these adjustment factors:

   1 meter ~ 1 yard + 10%
   1 kilogram ~ 2 pounds + 10%
   1 liter ~ 1 quart + 5%
   1 Celsius degree ~ 2 Fahrenheit degrees − 10%

The adjusted values are all within about 1 percent of the true values, which will almost always be good enough for making estimates.

Chapter 2

Millions, Billions, Zillions

"Perhaps the Bush administration could use the 660-billion-barrel
Strategic Petroleum Reserve to push prices down."
(Newsweek, May 24, 2004)

Some years ago, at a time when gasoline prices had risen
significantly (although they were still well under two dollars a
gallon), *Newsweek* suggested increasing the supply of gasoline
in the hope of reducing prices for consumers. The United
States maintains a large emergency reserve of petroleum in
underground salt caverns along the Gulf of Mexico in Texas
and Louisiana; the idea was that releasing some of that onto the
open market would increase the supply and thus drive down
prices.

Besides the size of the Reserve, the article provided another
useful fact: "The average vehicle uses 550 gallons per year."

Thus, a natural question might be to ask how long the Strategic Petroleum Reserve would last if it were just used to satisfy consumer demand. If you like, take a moment and try to work that out for yourself. You might begin by converting 550 gallons in a year to something more readily visualized: dividing 550 by 365, it's close to 1½ gallons a day.

2.1 How long will it last?

To figure that out, we need to know how many vehicles there are, and how big a barrel is.

How many vehicles are there? In the last chapter, we came up with 200 to 250 million. That's good enough for now and we can fix it up later if we learn more.

How big is a barrel? That's harder, but if we think about the 55-gallon drums that litter construction sites and dumps, or perhaps the beer kegs that one sometimes sees at parties or behind restaurants, we can make an informed guess. Since we don't know for sure, let's call a barrel 55 gallons, and come back later to adjust that if necessary.

One reason for assuming that a barrel is 55 gallons is that it makes the arithmetic easy. If each vehicle uses 550 gallons per year and a barrel is 55 gallons, then each vehicle uses 10 barrels per year. Multiplying 250 million vehicles by 10 barrels per vehicle gives 2.5 billion barrels per year total. This computation is a bit rough since we had to estimate both vehicle count and barrel size, but it's not likely to be wrong by a huge factor.

Newsweek says that the Reserve holds 660 billion barrels, and we use about 2.5 billion barrels per year. 660 divided by 2.5 is 264, which tells us that the Reserve should last over 260 years! Why are we so worried about oil? It sounds like we

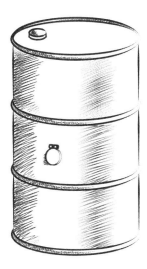

Figure 2.1: How big is a barrel?

could just ignore all the troubled parts of the world where war and politics interfere with oil production; the United States can tell them to go away, since it has plenty of oil already.

Something is wrong.

2.2 How can this be?

This is the place where if I'm talking to a group of people—giving a lecture, for example—someone raises an objection. The estimate of vehicles is too low because I didn't include trucks and buses, and they use a lot of fuel. Or it fails to take into account population growth. Or a barrel is smaller than I said. Or the refining process doesn't convert 100 percent of the crude oil to gasoline. Or there are other uses for petroleum.

All of these are perfectly valid points. But even if my estimate is wrong by a factor of two, three, or even ten, the conclusion remains the same: there's plenty of oil in the Reserve and it will last a long time. Something is *really* wrong here, not something that can be fixed by minor adjustments. What's going on?

The answer became clear a couple of weeks later, when *Newsweek* published a correction: "... we said that the size of the Strategic Petroleum Reserve is 660 billion barrels. It is actually 660 million barrels." In other words, *Newsweek*'s confusion between million and billion resulted in an error of a factor of a thousand, and that is a big deal.

We could repeat the arithmetic with millions instead of billions, but there's no need; we've already done the work. If we divide 250 years by 1,000, that's one quarter of a year. The Reserve would last only three months! Dipping into the Reserve would make at most a tiny temporary dent in gas prices, while using up the oil in almost no time. The country is right to be worried about oil, and the president was wise to ignore this suggestion.

As an aside, this idea resurfaces from time to time. President Obama considered it in 2011; on March 7 *Business Insider* said "the White House appears to be attempting to talk down the price of crude by referencing its 800 billion gallon reserve of oil." The same story referred to the same reserve as 727 million barrels (the correct value) only five short paragraphs later, which suggests that the article was glued together in haste and never read carefully.

Million and billion are confused surprisingly often: surprising because a billion is a thousand times bigger than a million, that is, a factor of 1,000. Let me try to make that factor more meaningful. Suppose you think you have $100 in your pocket right now. If that's too small by a factor of a thousand, then you really have $100,000, which is enough to buy a fancy car or even a modest condo in some parts of the country. On the other hand, if the amount is too big by a factor of a thousand, then you only have ten cents, and that's not enough to buy anything.

The *Newsweek* story is not all that atypical. A reliable and responsible source says something about big numbers. Others may act on the story, or pass it on. Most of us let it wash over us, leaving no trace except perhaps a vague feeling that somebody should do something. But as we just saw, an analysis that used nothing more than general knowledge, rough approximations, and a bit of elementary-school arithmetic revealed a major error in the story.

How many of the numbers that we see every day are equally wrong, a factor of a thousand off in one direction or the other, and just as misleading? And how many are produced by much less reliable and disinterested sources, whose goal is not to inform but to sell us some thing or some idea?

Take a look back at how we spotted the error. The first step, of course, was to pause long enough to think about it. Then we made some rough but defensible estimates of a couple of the values we needed. We did some simple arithmetic, and came up with a conclusion that couldn't possibly be right. No matter how rough our estimates and arithmetic, there's no way they could have been off by a factor of a thousand. Thus some number in the original story must have been wrong.

We'll see this same pattern throughout the rest of the book, as we explore how to spot potential problems, how to make reasonable estimates, how to do approximate arithmetic easily, and how to reason backwards from conclusions to truth or falsehood.

2.3 Check the units

Years ago, when I first started thinking about the general area of numeric self defense, oil prices were rising rapidly. They went even higher, fell back somewhat, then rose again, in a cycle that's likely to continue until we no longer depend on fossil fuels. Energy is an important topic today and will probably remain so for a long time, so there's no shortage of news stories that use big numbers to make their points about prices, environmental concerns, and the like.

It's easy to make mistakes when there are two units in play, especially when one of them is not familiar from everyday life. Thus, the *New York Times* said in an editorial on April 26, 2006, "The [strategic petroleum] reserve has a capacity of 727 million gallons"; on October 3, the *Times* corrected the units to barrels. The capacity number is ten percent higher than *Newsweek*'s number. The official web site at energy.gov says that the capacity is somewhat over 700 million (not billion!) barrels. These values, from different sources and different times, are all pretty close to each other; such consistency is a good sign.

How big is a barrel? It turns out that an oil barrel is smaller than the familiar 55-gallon drum; as the *New York Times* said on June 9, 2010, correcting a story about the oil spill in the Gulf of Mexico, "A barrel holds 42 gallons, not 42,000 gallons." (There's that factor of a thousand again.) We originally estimated that an oil barrel was 55 gallons, so our estimates were wrong, but only by 25 or 30 percent (42/55 is 0.76; 55/42 is 1.31), so it's not a big deal, especially when we don't know the other numbers accurately either.

The Gulf of Mexico oil spill mentioned above was caused when the Deepwater Horizon drilling rig exploded and sank in April 2010. The spill went on for three months before being brought under control; while it was active, it provided its own steady flow of numbers, often wrong and sometimes seriously misleading.

It was bad enough that the operator of the rig and various government agencies couldn't make accurate estimates of the volume of oil that was escaping, but the errors were compounded by erroneous units and factors. For example, a *New York Times* story in May 2010 said that the drilling rig was carrying 750,000 barrels of diesel fuel. Barrels were corrected to gallons shortly thereafter.

Oil is always in the news. On January 4, 2008, the *Newark Star-Ledger* said "a map showing oil production around the world [yesterday] misstated the amount of crude pumped each day. For each country listed, the amount should have been in millions and not billions of barrels."

In March 2008 the *New York Times* said that Americans used 3.395 billion gallons of gasoline in 2007. Since the population of the US is over 300 million, each American must have used 10 gallons during the year. This is clearly far too low; filling up a single car even once would use more than a person's annual amount. Replacing gallons by barrels brings it up to about 10 barrels per person per year, nearly the same number as we saw above. That consistency is a helpful check: if independent sources and computations come up with similar answers, they're more likely to be correct than if they are wildly different.

Later that year, we learned that Cuba's off-shore oil reserves contain "as many as 20 million barrels," which seems small enough that one might suspect (correctly) that it should have been 20 billion.

The *New York Times* said in April 2008 that Mexico's oil production in the previous year dropped to about 3.1 billion barrels a day. Since there are over 7 billion people on earth, that would imply a production of nearly half a barrel per person every day, and just from Mexico! If that were correct, we couldn't get rid of the stuff fast enough. Sure enough, there was a correction soon after, replacing billion by million. Gallons versus barrels (a factor of 42) and millions versus billions (1,000) are both common errors.

How does one spot such errors? It's helpful to know a few relevant facts. First, as we saw above, there are roughly 250 million vehicles in the US. Second, the national average is about 10,000 miles driven per year per car. Cars get about 20 miles per gallon, so annual usage is about 500 gallons, or somewhat over 10 barrels. If you know a couple of these numbers, you can estimate the others. For example, is 10,000 miles per year a reasonable value? As we saw in Chapter 1, if you drive 20 miles each way five days a week, that's 200 miles per week, and if you do that for 50 weeks, that's 10,000 miles. Of course this story would be different in detail in other parts of the world, for example in Europe, which has higher gas prices, shorter distances, and better public transportation.

Another common way to muddle things is to confuse time units. On February 12, 2007, the *Newark Star-Ledger* issued a correction: "In an editorial yesterday about alternative fuels, it was incorrectly stated that the use of motor fuel in the United States could rise to 170 billion gallons per day in 10 years. It should have said 170 billion gallons a year."

Reputable news outlets try hard to get their facts right, and they make a point of correcting errors prominently, for which they deserve credit. For instance, in May 2010, the *Wall Street Journal* published a correction that read "The euro zone consumed 10.5 million barrels of oil a day last year. A May 21 *Heard on the Street* article about the impact of Europe's woes on commodity prices incorrectly said the euro zone consumed 10.5 million barrels last year." Mixing up days and years leads to an error of a factor of 365.

2.4 Summary

Looking back at the examples in this chapter, we can see some ways to reason about numeric claims.

First, it's important to know some real facts—how many people there are in various parts of the world, how big everyday items are or how much they weigh, how often things happen, and the like. Real-world experience is a big help here, and the more experience you have, the more likely that relevant facts will be near to hand when you need them. The Internet is invaluable but it may not always be available, and even if it is, it may not be accurate.

Second, there's no need for precise arithmetic; ballpark figures and approximate computations are fine. Whether a barrel of oil is 55 gallons or 50 or 42 makes little difference if the fundamental error is a factor of a thousand, so we can safely cut corners, round values off to multiples of 5 or 10 for easy arithmetic, and generally take computational shortcuts. In practice, if one such approximation or estimate is too high, another might be too low, so the results will tend towards sensible values automatically.

Third, we can reason backwards from conclusions to assumptions and given data. If some number is claimed to be correct—for instance, the petroleum reserve will last 250 years—what are the implications of that? If the implications are nonsensical or plain impossible, that means there must be some mistake, and we can backtrack to figure out what might have gone wrong.

Fourth, we can look for consistency among independent computations or sources. If there are multiple ways to arrive at a value, the different values should be reasonably close to each other; if they are not, something is awry. We saw this with independent computations that all seemed to suggest that the average distance driven in a year is about 10,000 miles in the US. If one computation says the average is a thousand miles and another says a hundred thousand, at least one is wrong, if not both.

Finally, and most important, we can use our heads. Rather than taking numbers at face value, or letting them roll over us unexamined, we can think about whether they might make sense, or whether perhaps they're a bit flaky. With a little practice, this gets easier, and as a result we become more confident in making our own estimates and in assessing numbers from sources like newspapers, TV, advertisers, politicians, government agencies, bloggers, and random web sites.

Chapter 3

Big Numbers

"zillion: A generic word for a very large number.
The term has no well-defined mathematical meaning."

(Wolfram.com)

Words like million, billion and trillion have no intuitive meaning to most people, myself included, and thus we tend to treat them as synonyms for "big," "really big," and "really really big." Over the years, I've collected hundreds of examples where a newspaper story or magazine article has used one of these words when a different one was needed. Indeed, one can search online for variants of "millions, not billions" and get a long list of corrections; presumably plenty more have gone uncorrected, noticed by no one at all.

These "big number" words tend to show up in business and finance (big bucks), government (big budgets and big deficits), politics (big promises) and social concerns (big populations and big problems). In this chapter, we'll look at some examples, and talk about how to cut the big words down to size and put some meaning on them.

3.1 Number numb

In September 2008, at the height (or, more accurately, depth) of the US financial crisis, a blogger named T. J. Birkenmeier posted a story with an interesting idea, which he called "The Birk Economic Recovery Plan," quoted verbatim here:

> "I'm against the $85,000,000,000.00 bailout of AIG. Instead, I'm in favor of giving $85,000,000,000 to America in a "We Deserve it Dividend". To make the math simple, let's assume there are 200,000,000 bona fide U.S. Citizens 18+. Our population is about 301,000,000 +/− counting every man, woman and child. So 200,000,000 might be a fair stab at adults 18 and up. So divide 200 million adults 18+ into $85 billon that equals $425,000.00."

Birkenmeier's plan resonated with many who felt a sort of populist anger at the financial institutions that had failed so completely in their fiduciary duties, and the blog post went viral. Among the typical responses were comments like these (also verbatim):

> "Ordinary citizens with great common sense sometime make our lawmakers and economists look like they are still in first grade. Here's a great example!"

> "I REALLY like this plan!"

> "Interesting idea. Of course, the politicians would never do anything this logical."

> "THE PERFECT SOLUTION. Sounds reasonable, don't you think?"

> "i'd vote for that puppy tomorrow!"

"DUH! Seems like a no brainer to me!"

"Here's a GREAT IDEA!! I don't know who this Birk fella is but I would vote for him for president."

Other readers weighed in with a somewhat different assessment, based on their own computations:

"The real reason this wouldn't work is simply due to the math. 85 bil devided by 200 mil equals 4250."

"Whoever wrote this plan needs to buy a calculator...it is not anywhere near $42,500 per person...42,500 per adult for 200 million people would be 8.5 Trillion dollars."

A handful of commenters both read the post carefully and did the arithmetic correctly:

"Uh, do the math. Its only $425.00. You may resume smoking whatever it is you're smoking."

Birkenmeier himself clarified: it had all been an experiment, one that successfully made his point:

"I wanted to see how many people really do do the math. So I sent the message below to 100 of my pals at random. I wanted to see how many folks would catch my intentional three-digit error...just three little zeros. So far only 2 people have actually done the math and let me know about it. [...]

So what's my point? We are all number numb. And, very few people, even really smart folks rarely do the math."

The phrase "number numb" is a good description of the problem for many of us—there are too many numbers to assess, to the point where we ignore them or take them at face value, rather than thinking about them. Even when we do find time and inclination, arithmetic errors can compound the problem.

Since big numbers don't carry enough intuitive meaning, we need to cut them down to size, to bring them into a range where there's a decent chance of understanding them.

3.2 What's my share?

One good way to reduce large numbers like the national debt or the cost of a corporate takeover to human scale is by expressing them as the amount per person or per family. For example, on October 24, 2010, an editorial in the *New York Times* said "The yearly budget deficit stands at $1.3 billion." Converting billions to millions, $1.3 billion is $1,300 million. If there were 300 million Americans at the time, my share of the deficit, and yours if you lived in the US, was 1,300 million divided by 300 million, or 1,300 divided by 300, or a little over four dollars.

Accordingly, I offer not one but two plans for reducing or even eliminating the deficit. First, we could declare one day as "Do without fancy coffee day." Rather than having an expensive coffee and a muffin, each person instead sends four dollars to Washington. Deficit eliminated, in one painless day of minimal sacrifice.

Alternatively we might find some public-spirited billionaire, perhaps one of the bankers or hedge fund operators who made out so well during the financial crisis, who would be willing to simply pay off the deficit himself or herself.

What's wrong with my plan? This is a place where the "what would that mean to me personally?" approach applies. The number involved, $4, is so small that it's within range for an individual. But that makes no possible sense—if the deficit were that easily dealt with, it would have been—so something is wrong. It's trillions instead of billions this time, of course: the deficit was $1.3 *trillion*, which is $4,000 per person, and there's little hope that each of us will send that much money to Washington if we don't have to.

Let's try another example. If the US budget is $3.9 trillion (about right for 2016) and there are 300 million Americans then each person's share of the budget is $3.9T/$300M. It's easier to do arithmetic if the units are the same, so start by converting trillions into millions: $3.9 trillion is $3,900,000 million. Divide that by 300 million people, and we get $13,000 per person. If there are four people in a typical family, that's $52,000 per family.

Could you imagine paying your share all by yourself? In a sense, we all do, through some combination of personal and corporate taxes, though the burden falls very differently on different people. But at least the average seems within range, which is a good sign.

Of course if you want to do this kind of calculation, you have to know something about the number of people involved; that's one reason why it's a good idea to have ballpark figures in hand for how many people there are in the world (about 7.5 billion in 2017), your country (1.4 billion for China, 750 million for the European Union, 36 million for Canada), your state or province (40 million for California and 14 million for Ontario, for instance), and perhaps the town or city that you live in (30,000 for Princeton, 100,000 for Boulder, 800,000 for San Francisco, 9 million for London, 22 million for Beijing). All of these are approximations of moving targets, but they are good enough for reasoning about your individual share of budgets, deficits, taxes and the like for at least a few years in the past or into the future.

In August 2007, the *New York Times* stated that the average annual income for all Americans from 2000 to 2005 was $7.43 billion. Does this make sense? It would imply that the average income per family was $7 billion divided by say 100 million families, or around $70 per family, which even in perilous economic times is clearly nonsense. The original number should have been $7.43 trillion; that factor of 1,000 brings average family income up to $70,000, which seems high but is probably within a factor of two. (Chapter 9 explains why an arithmetic average like this may not be the best way to characterize income when there is a wide range of values.)

The United States is not the only place in the world that has had financial problems, of course; the European Union has had to bail out several weak economies. The same kinds of big financial numbers and big errors appear; as the *New York Times* reported on May 25, 2010, the rescue package was "750 billion euros, not 750 million euros." The population of the European Union is around 750 million; if the rescue really only required one euro per person, it wouldn't have been hard to arrange.

And as one final example, a web site that I visited recently said "Americans spend $1.7 billion annually handling chronic diseases." It looks like our share of the chronic-disease bill is about $5 per year for each of us, doesn't it? The cost of dealing with chronic conditions is a significant part of health care costs in most countries, and certainly in the US where there is an endless acrimonious and ideological debate about how health care should be funded. If the cost were only five or ten dollars per person per year, perhaps the debate would be easier to resolve. Unfortunately, the spending is $1.7 trillion, not billion. If each person's share is thus five or ten thousand dollars a year, that's certainly big enough to warrant debate, and you can see how it relates to annual income and annual taxes, at least approximately.

3.3 High finance

Big numbers abound in the financial industry, where companies sell billions of dollars worth of goods and services each year, and are themselves bought and sold for billions as well. Exalted beings are also so wealthy that their worth is measured in billions—the 2017 *Forbes* list of the world's billionaires has 2,043 names! (That's up from 937 in 2010.) When I last looked, Jeff Bezos, founder of Amazon, was worth nearly $130 billion, Bill Gates was worth $90 billion, and Warren Buffett around $84 billion; they are likely to be even better off by now.

I'm in no immediate danger of becoming a billionaire or indeed anything remotely close to it, and I suspect you aren't either, but the same kind of "what would that mean to me personally" reasoning can be used to assess big financial numbers, even if few of us are ever likely to experience such off-scale wealth.

An article in the *New York Times* in May 2006 said that the sale of *The Philadelphia Inquirer* and *Daily News* newspapers was expected to bring $600,000. I couldn't pay that much myself, but that price is clearly within range of one perfectly ordinary person, and there would be some cachet in being able to say, "Oh, yes, I own the *Philadelphia Inquirer*," a respected paper that began publishing in 1829. As you might suspect by now, the number should have been $600 million, dashing my hopes of a personal media empire.

The "On Media" column on the *MediaDailyNews* web site in February 2008 said that MySpace's estimated worth was

$10 million, out of reach for most individuals but manageable by a small consortium of well-off friends. The corrected figure was $10 billion. Ask yourself "Could I afford it personally?" That's often a valuable corrective.

Of course, MySpace went through some bad times soon after, and actually did trade hands for $35 million only a few years later, so perhaps the original article was prescient, instead of wrong by a factor of a thousand.

A 2005 article in *Business Day* about Verizon's stock price said that the amount that Vodafone was likely to want for its 45 percent stake in Verizon Wireless was $20 million. This is consistent with an article in 2008 that gave Verizon's 2007 revenues at $93.4 million. Although consistency is almost always a good thing when reasoning about numbers, unfortunately both of these dollar figures should have been in billions, not millions.

The flip side of a ridiculously low value for a big company that you've heard of is a very large value for a company that you've never heard of. For instance, a March 2010 *Associated Press* story about Sonic Corporation (who?) in the *Seattle Times* said that the company's revenue totaled $112.8 billion, which was significantly larger than the revenues of a couple of other Seattle-area companies that you might have encountered: Microsoft and Amazon. A subsequent correction scaled Sonic's revenue back to millions.

Closer to home, a 2008 story in my local paper reported that a nearby veterinary practice's "estimated annual billing was $18 million, not $18 billion." It caught my eye because we had in fact once taken our cat there for treatment; it did not look like a multi-billion dollar enterprise, though they did charge us a hefty sum.

3.4 Other big numbers

Not all big numbers involve money. For instance, in March 2008, the *New York Times* said "the number of people [in India] who rely on animal waste and firewood as fuel for cooking [...] is about 700 million, not 700,000." Both of these seem surprising to someone with only a superficial knowledge of India; one might have expected something between.

Around the same time, a newspaper story "misstated the number of Catholics in South America. It is 324 million, not 324,000." The smaller number would seem unlikely, given that much of South America was settled by people from Spain and Portugal, both strongly Catholic countries.

The physical world is the source of many numbers large and small, and thus provides another place for things to go wrong. It's helpful here to know some real facts, like the age of the universe (about 14 billion years), distances to the moon (240,000 miles or 380,000 km), to the sun (93 million miles or 150 million km), around the world (25,000 miles or 40,000 km), and across the country. The speed of light (186,000 miles or 300,000 km per second) and the speed of sound (1,120 feet/sec or 340 m/sec) are also useful values to have in mind.

"Scientists now say the Big Bang happened 13.7 billion years ago, plus or minus 150 million years—not plus or minus 150,000 years." As the *San Francisco Chronicle* said in January 2006 in an article about two stars in the Milky Way galaxy, the age of one of the stars is 300 million years, not 300 billion years. If you know a few of the key values, you can more easily detect places where something is wrong by a big factor.

One caveat: reasoning backwards and scaling up or down are invaluable tools, but they can't identify all numeric problems. For example, on February 27, 2018, the *New York Times* printed a correction: "An article on Sunday about Warren Buffett's annual letter to Berkshire Hathaway's shareholders misstated the book value for Berkshire Hathaway in 2017. The value rose to $348 billion, not $358 billion." Although the dollar amount is large, the percentage error (less than 3 percent) is so small that no casual reader could spot it; fortunately, the *Times* is scrupulous about fixing even minor misstatements.

3.5 Visualizations and graphical explanations

Journalists are fond of visual imagery to try to convey an impression of size or scale, like this example from the *New York Times* in August 2000, describing a massive recall of defective car tires: "If the 6.5 million Firestone tires recalled so far were stacked vertically, they would make a column 949 miles high."

Is this computation accurate? We can figure that out. If the tires are stacked on their sides and each tire is one foot wide, a stack of 6.5 million tires will be 6.5 million feet high. We could divide 6.5 million by 5,280 (feet in a mile) or simplify by dividing 6 million by 5,000; both lead to about 1,200 miles. If instead the tires are 9 inches wide, that's ¾ of 1,200, or 900 miles. So the computation is correct, though excessively precise, a topic that we will come back to in Chapter 8.

I'm not convinced that visualizations like this one are helpful, except to convey the impression that some number is "big" or perhaps "really big." After all, what's your mental picture of 949 miles, especially straight up? (Check Figure 3.1.)

A visualization that brings the data into a range where it's easy to relate to is a different story. Instead of an improbable tower of tires, we could say "There are 330 million people and

Figure 3.1: Stacks of tires.

6.5 million recalled tires, so that's one tire for every 50 people." That's easier to visualize—if we're in a place like a bus or a store or a classroom with 50 people, one of them would have had a tire recalled.

Visualizations are based on the assumption that an image is familiar to the audience, which is not always the case. A TV news story described a ship as being "nearly three and a half football fields long," a parochial image that might not be helpful outside the US, and which could have been better expressed as "nearly 350 yards (320 meters) long."

Football analogies are popular in the US. An article on the perils of texting while driving reports that "motorists who send or receive a text message have a tendency to take their eyes off the road for five seconds to do so. That is enough time for their car to travel more than the length of a football field at highway speeds." If you don't know how big a football field is, you have no way of knowing whether this is important or not. (Of course taking your eyes off the road for five seconds is likely to be a bad idea no matter how big the field is.) Or we might say that since "football" in much of the world is the game Americans know as soccer, and a soccer field is only moderately bigger, no harm has been done.

3.6 Summary

As Daniel Kahneman, winner of the 2002 Nobel Prize in economics, and author of *Thinking, Fast and Slow*, once said,

> "Human beings cannot comprehend very large or very small numbers. It would be useful for us to acknowledge that fact."

One of the most effective ways to understand big numbers is to try to scale them down, for example, by asking what your share of a big number is or how it would affect your family or some other small group. No one can personally relate to a trillion dollar budget, but it's reasonably intuitive to say that your personal share of the budget is a bit over $3,000.

Visualizations of big numbers are a mixed lot. Some work well, but in many cases they simply replace an unintuitive number by a similarly unintuitive image, like a pile of tires or a trip to the moon. And they are less helpful if they're based on a cultural reference like football fields that doesn't translate well from its origin.

Chapter 4

Mega, Giga, Tera, and Beyond

"A zettabyte is equal to one billion trillion bytes: a 1 with 21 zeros at the end. A single zettabyte is equivalent to 100 billion copies of all the books in the Library of Congress."

(New York Times, December 10, 2009)

Technology is a plentiful source of big numbers, many expressed in unfamiliar units, so there is another set of "big" words to add to the mix: mega, giga and tera are part of everyday speech, while further-out ones like peta and exa now appear in public with some regularity. Computers and smartphones are so pervasive that we're all used to reading about gigabytes and megapixels, but since these prefixes often refer to invisible entities like bytes, we have even less idea of what these terms mean than the more familiar billions and trillions.

To put everyone on the same footing, kilo is one thousand, mega is one million, giga (pronounced with a hard "g" as in "gig") is one billion, and tera is one trillion. If you want to future-proof yourself as technology advances, the rest of the sequence is peta, exa, zetta and yotta. Each one is 1,000 times bigger than the previous one.

In a similar vein, computers are so fast and made of such tiny components that there is a parallel universe of even more unfamiliar prefixes for small quantities and sizes: milli, micro, nano, and pico, which are one thousandth, one millionth, one billionth, and one trillionth. These are most often applied to lengths and times, like millimeters and nanoseconds.

Since most of us have little intuition about such numbers and don't know the data that they are based on anyway, we're at the mercy of whoever provides them. Here are several illuminating examples.

4.1 How big is an e-book?

There was much pre-Christmas buzz some years ago about Amazon's Kindle and other newly-available e-readers as potential gifts, along with speculation about a tablet device from Apple. (The iPad was announced in late January of 2010, but didn't ship until March.) On December 9, 2009, the *Wall Street Journal* said that Barnes & Noble's Nook e-book reader had two gigabytes of memory, "enough to hold about 1,500 digital books." A day later, the *New York Times* made its observation that a zettabyte "is equivalent to 100 billion copies of all the books in the Library of Congress."

By good luck, I was right then in the early stages of inventing questions for the final exam in my class, so this confluence of technological numbers was a gift from the gods. On the exam, I asked, "Supposing that these two statements are correct, compute roughly how many books are in the Library of Congress."

This required only straightforward arithmetic, albeit with big numbers, which is not something that most people are good at. The brain refuses to cooperate when there are too many zeroes. Writing them all out ("a 1 with 21 zeros at the end") might help, but it's easy to slip up. As we will see shortly, scientific notation like 10^{21} is better, but units like zetta, completely unknown outside a tiny population, convey nothing at all to most people.

Since intuition is of no help here, let's do careful arithmetic. Taking the *Journal* at its word, 2 gigabytes (2 billion bytes) for 1,500 books means that a single book is somewhat over a million bytes. Taking the *Times* at its word, a hundred billion copies is 10^{11} copies; dividing 10^{21} total bytes by 10^{11} copies implies that there are about 10^{10} bytes in a single copy of all the books. If each book is 10^6 bytes, then (dividing 10^{10} by 10^6), we conclude that the Library of Congress must hold about 10^4 or 10,000 books. (If the use of exponents and scientific notation here is unfamiliar, there's more explanation in the next section.)

Is 10,000 books a reasonable estimate? One useful alternative to blind guessing is a kind of numeric triage, which led to the second part of the exam question: "Does your computed number seem much too high, much too low, or about right, and why do you say so?" Of course if one didn't do the arithmetic correctly, all bets are off. A number of students found themselves in that predicament, and thus had to rationalize faulty values from fractions to bazillions.

Figure 4.1: 10,000 books? One building of the Library of Congress.

Those who did the arithmetic correctly were better off, but some still had trouble assessing plausibility. Apparently even small big numbers are hard to visualize, for a surprising number of students thought that 10,000 books was reasonable for a big library. "I would guess that even Princeton's library holds over 10,000 books" was one response. That's technically true, of course, but not a good answer—even I have over 500 books in my office, and I'll bet that many of my more scholarly colleagues have thousands. The university library, housed in a large building at the center of campus and ostensibly familiar to all students, holds well over six million.

How big is a book really? How big is a terabyte, or even a megabyte for that matter? Here's a partial answer. One byte holds a single alphabetic character in the most common representation of text. Jane Austen's *Pride and Prejudice* has about 97,000 words or 550,000 characters, so a megabyte for a purely textual book like a romance novel or a biography is a good round number, and a gigabyte could hold a thousand books of similar size. (Pictures take up more space, kilobytes to megabytes each.) The *Journal*'s computations were reasonable, but the *Times*, by contrast, was way off.

Now we can assess these three quotations about the sizes of e-books:

> "If you had a word processing file holding the King James Bible, it likely would consume somewhat less than 500 kilobytes."

> "In terms of text, each gigabyte holds the equivalent of 2,000 Bible-size books."

> "Entire programs like Microsoft's Office suite take up about 1 fat book's worth of drive space. Microsoft's Office Small Business Edition, for example, takes up a mere 560 MB."

The Bible is quite a bit longer than *Pride and Prejudice*; at almost 800,000 words or about 4.5 megabytes of plain text, it qualifies as a fat book. The first two of these statements are thus reasonably consistent, though optimistic. (Data compression techniques can reduce the number of bytes required, though not quite down to 500 kilobytes.) The third is off by a factor of a thousand, however, since 560 MB of Microsoft Office is more like 500 fat books worth of disk space.

By the way, according to loc.gov, the Library of Congress has more like 16 million books and 120 million other items. As an amusing aside, the original story about e-book readers tried to help readers to visualize how many books there are in the Library of Congress: it would be "seven layers of textbooks covering the continental United States and Alaska." I'll leave it to you to decide whether this is accurate (passing over whether it's helpful). To give you a start, however, consider that a square mile is well over 25 million square feet, and a textbook isn't much bigger than what you're holding as you read.

4.2 Scientific notation

When they get to numbers that are beyond "really really big," news sources resort to compounding. As the *New York Times* said in a correction in March 2008, "[A petaflop] can be expressed as a thousand trillion instructions per second, not a million trillion." Or, quoting *Computerworld* in December 2007, "In the private sector alone electronic archives will take up 27,000 petabytes (27 billion gigabytes) by 2010." Or, in June 2017,

> "According to CERN, the long-sought [Higgs] boson, the keystone to the Standard Model, weighs 125 billion electron volts, or as much as a whole iodine atom. But that is ridiculously too light, according to theoretical calculations. The mass of the Higgs should be some thousands of quadrillion times as high."

Not only compounds, but mixtures of regular big numbers like billion and trillion, rarer ones like quadrillion, and their technology siblings giga and peta! What is the poor reader to do?

One way to deal with big numbers is to write them out in their full glory, rather than using words like million and billion.

So a million is 1,000,000, and a billion is 1,000,000,000. Much further beyond, as the *Times* explained, "A zettabyte is equal to one billion trillion bytes: a 1 with 21 zeros at the end." (The "21" comes from adding the 9 zeroes for a billion and the 12 zeroes for a trillion.)

In *scientific notation*, we use a power of 10 to express the number of zeroes that follow the 1. In this notation, a thousand is 10^3, that is, 10 to the third power, or 10 multiplied by itself 3 times ($10 \times 10 \times 10$). Similarly, a million is 10^6, billion is 10^9, and a trillion is 10^{12}, that is, 10 to the 12th power, or 10 multiplied by itself 12 times. To multiply powers of 10, like 10^9 times 10^{12}, add the exponents: 10^{9+12} is 10^{21}. For division, subtract the powers: 10^{21} divided by 10^{11} is 10^{21-11} or 10^{10}.

This is simple, compact, and less vulnerable to errors than using big words or counting zeroes. For example, in *Broadbandits*, a 2003 exposé of the telecom industry meltdown, the author says that a data transfer rate of 6.5 terabits per second is "almost a million times faster" than 56 kilobits per second. Is the factor of a million correct? Comparing 6 terabits (6 times 10^{12}) to 60 kilobits (60 times 10^3, which is 6 times 10^4), it's easy to see that the correct factor is close to 10^8, or 100 million times faster.

Unfortunately, however, many people are uncomfortable with scientific notation, so it doesn't get used as often as it could be in day-to-day life.

Sometimes technology gets in the way of clear expression: newspapers seem to be unable to print superscripts. A story in the *New York Times* in December 2007 said that chess is a much harder game than checkers for computers to master because it has between 1040 and 1050 possible arrangements of pieces, while checkers has more like 1020 positions. That doesn't sound like much of a difference, does it? But display these the way they should have been, and it's much clearer: chess has 10^{40} to 10^{50} arrangements, while checkers has 10^{20}. The difference between checkers and chess can now easily be seen: chess is harder by a *factor* somewhere between 10^{20} and 10^{30}, which is (as they say), a 1 with 20 or 30 zeroes after it: 1,000,000,000,000,000,000,000,000,000,000. Clear enough?

These factors are large. Suppose a computer could evaluate a billion (10^9) chess positions per second—that's fast for today's home computers but not for supercomputers. There are 86,000 seconds in a day or about 30 million (30×10^6) seconds in a year. If the computer could check 10^9 times 30×10^6 or 3×10^{16} positions in a year, it would take 3,000 years to evaluate 10^{20} positions; 10^{30} positions would take 10 billion (10^{10}) times longer.

4.3 Mangled units

> "The amount of clenbuterol in the horse's system [...] was 41 picograms, not "petragrams." A picogram is one-trillionth of a gram; there is no petragram."
>
> (*New York Times* horse doping story, August 6, 2008)

Some units are so unfamiliar, or (like technology sizes) sound similar enough, that it's easy to inadvertently mangle their names and thus add to potential confusion. At Christmas one year, my wife gave me a copy of *Googled: The End of the World As We Know It*, by Ken Auletta. It's an engaging history and assessment of one of the most successful technology companies of the past few decades. The very last sentence, however, says that Google stores "two dozen or so tetabits (about twenty-four quadrillion bits) of data."

As the *Times* might have said, there is no tetabit; if quadrillion (10^{15}) is correct, then the word should have been petabits, since peta is 10^{15}. This led me to another exam question: "Assuming that the word should have been petabits, how many gigabytes does Google store?" Answering it required converting petabits to gigabits, then converting bits to bytes (by dividing by 8 bits per byte) to get 3 million gigabytes. But "tetabit" is also only one letter away from another valid unit, terabit, so the second half of the question asked, "If tetabits really should have been terabits, how many gigabytes would there be?" I'll leave that as an easy exercise.

As an aside, *Googled* was published in 2009. Technology marches on very quickly, and it won't be long before storage numbers are denominated in exabytes, and we'll undoubtedly see frequent stories about "a 1 followed by 18 zeroes."

4.4 Summary

The prefixes that stand for big and little numbers from technology—mega, giga, nano, and the like—have the same problems as conventional big words like million and billion: they don't give an intuitive feel for size, just a vague impression of relative levels of bigness. At the same time, they are less familiar, so the impression is even less likely to convey an accurate meaning.

Familiarity with the words will help, and becoming familiar and comfortable with scientific notation, the use of exponents instead of words or long strings of zeroes, will make all of them more comprehensible and meaningful. When you see compound strings like "million million trillion," take the time to convert them into exponents; it's easier to get an accurate impression of size and much easier to do computations with big numbers.

Chapter 5

Units

"Americans receive almost two million tons of junk mail daily."
(*Dear Abby* newspaper advice column, January, 1996)

Even with the advent of the paperless society and metaphorical tons of email spam, I still get plenty of physical junk mail at my home every day. But two million tons a day sure sounds like a lot. Is *Dear Abby*'s claim reasonable? Let's think about whether it is or not.

5.1 Get the units right

We can start by asking the question from Chapter 3: how would that affect me personally? Two million tons is four billion pounds. If there were 300 million Americans in 1996, that's over 13 pounds of junk mail per person per day.

Somehow that doesn't seem realistic, especially when I think of Joe, the long-serving and conscientious mailman who has faithfully delivered mail to my house for nearly twenty years. That would be 26 pounds every day just for me and my wife. Abby's value is clearly too high.

This seems likely to be a classic error of using the wrong units—gallons instead of barrels, kilometers instead of meters, seconds instead of minutes, days instead of months or years. The specific number might well be right, but if the wrong unit has been attached to it, the ultimate value is wrong.

I suspect that's the problem here: either the wrong time unit or an incorrect weight unit. For example, suppose "two million tons" is right but "per day" should have been "per month" or "per year." Thirteen pounds a month would be about six or seven ounces a day, which still seems high. But thirteen pounds a year would be more like half an ounce per person per day, or an ounce for a family of two. That might be a little low, but it's not unreasonable.

Alternatively, perhaps "two million" is right but "tons" should have been "pounds." Two million pounds is 32 million ounces; dividing that by 300 million gives us about 1/10 of an ounce per day per person. That seems low, though not absurdly, so it's a legitimate possibility. And I'm sure there are other options as well.

Think about how we reasoned through this. Convert the big number in the original statement into a smaller number that represents its individual effect on us. If that number is clearly wrong, think about what might have been wrong with the original statement, and work through some of the possible errors to see whether a simple change could explain the original and thus lead to a more likely answer.

5.2 Reasoning backwards

Reasoning backwards from a conclusion to check its data and assumptions is a valuable technique, applicable in many situations. Let's take a look at some more examples.

> "Shutting down your computer and monitor overnight rather than running them 24 hours a day will save $88 a day."
>
> (*Newark Star-Ledger*, December 2004)

This story was published at a time when CRT monitors were by far the most common form of computer display. It certainly sounds like turning off the monitor would not be just a good idea, but necessary for financial solvency. If it really did cost $88 for half a day of electricity use, there would be precious few personal computers at all, since the cost for electricity alone would add up to more than $30,000 per year. Even back in 2004, the number couldn't possibly be right.

If you know how much electricity costs, which is about 10-15 cents per kilowatt hour in my part of the world, and how much power a computer and a monitor use (typically 100-200 watts, or 1 or 2 tenths of a kilowatt, about the same as a couple of incandescent light bulbs), then you can estimate that it would cost a cent or two per hour to run a monitor. Running a computer and monitor 10 hours a day for a full year would cost about $80. That strongly suggests that the time unit in the original story should have been "per year," not "per day," and indeed that was the case, as the *Star-Ledger* made clear in a correction a few days later.

A story in the *London Times* in November 2004 described a NASA jet that can travel 850 miles in 10 seconds, or 7,000 miles per hour. The first number, 850 miles in 10 seconds, is clearly inconsistent with 7,000 mph: at 850 miles in 10 seconds, the jet would travel 5,000 miles in a minute and thus 300,000 miles in an hour. The story went on to say that an aircraft that can fly at ten times the speed of sound was about to be tested over the Pacific Ocean, with the possible goal of a "hypersonic" cruise missile that could travel from Los Angeles to Pyongyang in less than an hour. The speed of sound is somewhat over 700 mph, so 7,000 mph is perfectly plausible for a missile.

By the way, you perhaps learned as a child to estimate how far away a lightning strike was during a thunderstorm: every five seconds between the flash and the sound is a mile away. That works because 720 miles per hour is 12 miles per minute and thus one mile is 5 seconds.

What are the implications of a passenger jet that could travel 850 miles in 10 seconds? It would certainly reduce the pain of air travel. Imagine yourself getting on a plane in London. The plane takes off, and 40 seconds later comes an announcement: "Please fasten your seatbelts. We'll be landing in New York in a few minutes."

A speed of 7,000 miles per hour may be reasonable for a missile, but it's far too high for a commercial plane. Commercial jets fly at 500 to 600 miles per hour, and the Concorde's top speed was 1,300 miles per hour, nearly twice the speed of sound.

That makes one wonder about a story in the *Manchester Guardian* in September 1993 that said "[A Boeing 747] is a vehicle, a man-made object capable of hurtling down a runway at speeds well over 2,000 miles an hour and lifting off into the air." A 747 taking off is truly impressive but its takeoff speed is more like 200 miles per hour.

Figure 5.1: 2,000 miles per hour?

Meanwhile, back on the ground, in 2005 New York City tried to sell the aging Willis Avenue Bridge that crosses the Harlem River between Manhattan and the Bronx. The price was a mere one dollar, and the city would even deliver it free to any location within 15 miles. Sadly, there were no takers, so it was ultimately demolished.

A newspaper story at the time included a traffic analysis: the bridge was used by only 75,000 vehicles a year. A bit of arithmetic says that's about 200 vehicles a day or fewer than one every 5 minutes, which is lightly traveled indeed, especially in New York, a city with a population of at least 8 million. Not surprisingly, there was a correction a few days later: bridge traffic was 75,000 vehicles per day, not per year.

Some units are more technical and less familiar, and thus easier to get wrong. For instance, I once saw an article that described the use of electroshock therapy on children with severe behavioral problems. The story said that children were treated with current ranging from 15 to 45 amperes. Don't try this at home! The right units should have been milliamperes, or thousandths of an ampere; as Wikipedia says, 30 milliamps is enough to cause fibrillation and 30 amps would be instantly fatal.

And on a happier note, the *Newark Star-Ledger* ran a story some years ago about a local bar, Tierney's Tavern in Montclair, which was offering pitchers of beer for $1.25 during happy hour. A pitcher of beer is usually 60 ounces, or a couple of liters; that much beer would certainly make me happy, at least for a while. The story was later corrected. The price should have been $1.25 for a 16-ounce pint, not a pitcher—still pretty cheap, but roughly four times higher than the original.

5.3 Summary

It's just as easy to make a mistake with units as it is with millions and billions. Depending on the particular error, the effect might be smaller—confusing day with year is "only" a factor of 365—while mixing up pounds and tons is a factor of 2,000, and interchanging feet and miles is a factor of 5,280. ("The distance above sea level of Colorado Springs is 6,000 feet, not 6,000 miles.")

From time to time, a units error has serious consequences. In 1999, the Mars Orbiter space probe disintegrated in the atmosphere of Mars. The cause: one part of the software used data in conventional English units, while a different part used standard metric units. The discrepancy led to an invalid computation of the amount of thrust to be applied for an orbital correction, which brought the craft too close to the surface.

As another example, in 1983 an Air Canada flight ran out of fuel since the amount of fuel loaded had been measured in pounds when it should have been in kilograms; as a result, the plane had less than half the necessary amount. Because of faulty instrumentation and some human errors, this was not noticed until the engines stopped while the plane was flying at 12,500 meters somewhere over Manitoba. Thanks to good luck and some truly exceptional piloting, the plane landed safely, without power and most of its instruments, on a landing strip at an unused former air force base.

In some cases, errors in units can be spotted by reasoning backwards. In others, like the two incidents described above, there is no real solution except to be very careful.

Chapter 6

Dimensionality

"Young males can roam 60 to 100 square miles looking
for food and mates, but females stay close to the cave,
foraging within a 10-mile radius."

(*Newark Star-Ledger*, July 9, 1999)

It's clear that male bears wander around a lot of territory.
How does the area that male bears cover compare to the area
covered by the more stay-at-home females?

Let's do the arithmetic. A circle of radius r has an area of
πr^2, and π is approximately 3.14, so a circle with a ten-mile
radius has an area of over 300 square miles! Something is
clearly wrong, at least if we take at face value the implication
that lady bears stay closer to their caves than their gentlemen
friends do.

What might the problem be?

6.1 Square feet and feet square

It's easy to make errors when we mix a linear dimension like a radius in miles with an area dimension like square miles.

We've all heard from childhood that you can't add apples and oranges. Many of the numbers that we deal with have dimensions—length, area, volume—that must be combined correctly or we're doing the equivalent of adding apples and oranges. We can't add feet to square feet, nor compare square inches to cubic inches.

Fortunately, such errors are usually easy to spot. For example, a correction from the *New York Times* in May 2009 described a room as "30 feet square—meaning 30 feet by 30 feet—and not 30 square feet, which would be far less."

Indeed it would: a 30-square-foot room would be something like 6 feet long and 5 feet wide. Confusing "square feet" with "feet square" is common in English, where a bit of sloppiness in expression lets it in. A few months before the bear story, we learned that "the size of Fort Leavenworth is 8.8 square miles, not eight miles by eight miles, or 64 square miles." Or in the book *Alias Jack the Ripper* (R. M. Gordon, 2000), "The victims all lived in the same small area of 260 square yards." An area of 260 square yards would be about 16 yards by 16 yards. The author presumably meant 260 yards by 260 yards.

Reasoning backwards about the claimed area will often reveal such problems. During a flare-up of the Zika virus in Miami in the summer of 2016, newspaper stories quoted the director of the Centers for Disease Control and Prevention, Tom Frieden, as saying he would not be surprised if new Zika cases were reported inside of what he described as a 500-square-foot area at the center of the designated zone. "That's the way Zika works," he said, explaining that a one-square-mile zone was a precautionary buffer.

How big is 500 square feet? By coincidence, the room that I was sitting in when I first wrote this section was about 20 feet by 20 feet. That's 400 square feet; if the room were 22 by 23, that's 506 square feet. So if I had been in Miami instead of New Jersey, it's possible that new cases of Zika would have been localized to a place no bigger than my room! Surely that would have made the virus easier to control.

Obviously what Dr Frieden meant to say (and probably did say) was not "500 square feet" but "500 feet square," that is, an area of 500 feet by 500 feet, which is 250,000 square feet. The precautionary buffer of one square mile is the same as a square of one mile by one mile, so there's no error there.

6.2 Area

"The F50fd's sensor is more than 50 percent bigger than those on most of the other cameras: 0.625 inch diagonal, versus 0.4. Now that's a statistic—not megapixels—that matters in a camera."
(*New York Times*, December 6, 2007)

Many of the displays that we deal with in everyday life, like TV, computers and mobile phones, express their screen size as a single number, the diagonal measurement of a rectangular surface. That's convenient, since a single number characterizes the size, and it works well if the aspect ratio (the ratio of width to height) is the same across the devices that are being compared.

Camera sensors fall into the same category, though we don't see them. A sensor inside a digital camera is an array of millions of tiny light-sensitive picture elements ("megapixels") that measure incoming light and capture the values for later display.

The quotation above is absolutely correct in saying that a bigger sensor does a better job, since it gathers more light, but it's wrong in its arithmetic, at least if "50 percent bigger" refers to the sensor area. For a fixed aspect ratio, a 50 percent increase in the diagonal is also a 50 percent increase in both width and height, and thus the area increases by a factor of 2.25.

How did I reach that conclusion? The area is height times width, so if the old area was say h times w, the new area is 1.5 h times 1.5 w, which is 2.25 times the original. Another way to say it is that the area is bigger by 125 percent: if the old area is 100 square units, the new area is 225 square units. (This difference between factor and percentage is a potential confusion as well, so be careful.)

This might be easier to see with a diagram. The white squares in Figure 6.1 are the original ones; the gray squares are the ones added by growing both height and width by 50 percent. That's one more square in each direction, taking us from 4 squares to 9. And 9/4 is 2.25, which is a 125 percent increase.

This is exactly the same if the aspect ratio is different, that is, not square, as we can see in Figure 6.2. Indeed this is true for any shape, not just rectangles.

The author of the quotation does deserve full marks for rounding the increase of 0.625 over 0.4 to 50 percent to make it easy for readers to grasp. The ratio is really 1.5625, however, so the bigger sensor is actually 1.5625 squared, or 2.44 times as big—even better for camera buyers.

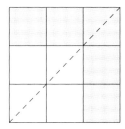

Figure 6.1: A 50 percent increase in the diagonal.

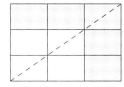

Figure 6.2: Also a 50 percent increase in the diagonal.

For consumers, the place where we see this kind of computation most often is television screens. At home, I make do with an ancient TV with a 38-inch (diagonal) screen, but every so often I think about upgrading. To make the arithmetic easy, let's suppose that I decide to go from a 40-inch diagonal to 60 inches. There's that 50 percent increase again, a factor of 1.5, so the screen area of my hypothetical new TV would go up by a factor of 2.25. If I had more interest in watching TV (and more money!), I could get an 80-inch set instead; that would give me four times the area.

Of course my big new TV is likely to have exactly the same number of pixels as my original small one unless I spring for "ultra HD," which has four times as many pixels. Why four times? That's because ultra HD has twice as many pixels in both height and width.

> "Ultra HD refers to a resolution of $3,840 \times 2,160$ pixels.
> That's four times the $1,920 \times 1,080$ pixels found in your
> full HD TV."
> (Product-comparison web site)

As it is for TV screens, so it is for computer displays—a 15-inch screen has 33 percent more area than a 13-inch screen, and 85 percent more area than my laptop with its 11-inch screen, assuming the aspect ratios are the same. Again, the pixel density may well vary too, so we have to be careful about what we're comparing.

6.3 Volume

> "A cannon with a bore diameter of 3 inches would shoot
> a slightly smaller iron ball that weighed between 3 and
> 4 pounds. A cannon with a bore diameter of 9 inches
> would shoot an iron ball that weighed between 7 and 10
> pounds. The size of the cannon came to be referred to
> in terms of the average weight of the solid ball shot they
> could fire. A cannon with a bore diameter of 3 inches
> was called a 3-pounder, one with a bore diameter of 6
> inches was called a 6-pounder, and so on."
> (From a web site on the history of artillery)

As we saw above, it's easy to make a major error by using a linear dimension like length or radius when an area dimension like square feet is needed, or vice versa. And if such errors are

significant for area, they are even more significant for volume.
This is easiest to see with blocks, as in Figure 6.3.

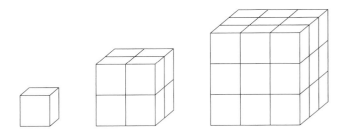

Figure 6.3: Growth of area and volume.

The area of a side grows as the square of the number of
blocks horizontally: for 1, 2 and 3 blocks, the areas are 1, 4 and
9. Similarly, the volume grows as the cube: 1, 8 and 27.

Now let's look at the cannonballs more carefully (Figure
6.4). The area of a circle of radius r is πr^2, a familiar formula.
If we double the radius, the area goes up by a factor of four.
The volume of a sphere like a cannonball is $(4/3)\pi r^3$, a formula that is less likely to be familiar (and that you can completely ignore). For comparisons, all that matters is that the
volume and thus the weight is proportional to the *cube* of the
radius or (equivalently) diameter. That means that if we double the radius or diameter, the volume and weight go up by a
factor of 8, and that's a lot. If a 3-inch cannonball weighs 3
pounds, then a 6-inch cannonball would weigh 24 pounds, and
a 9-inch cannonball would weigh 81 pounds.

The important thing to remember is how the volume (and
thus weight) grows in proportion to the linear dimension; constant factors like 4/3 and π don't matter.

Figure 6.4: 6-inch cannonballs for a 24-pounder.

You may remember a time when TV sets and computer monitors were not flat screens but had significant depth as well as height and width. If you wanted to replace an old 20-inch TV by a newer one with a 30-inch screen, the screen area went up by a factor of 2.25, but the volume of the set would go up by 1.5 cubed, which is a factor of about 3.3. I don't recall now, but would not be surprised to find out that the weight went up by some similar proportion, or at least more than a factor of 2.25. But with flat screens, the depth is constant and so the weight is likely to grow proportionally to area only: if my 40-inch TV weighs 10 kilograms, my prospective 60-inch flat-screen TV will weigh around 25 kilos. A bit of online comparison shopping suggests that this factor is reasonably accurate.

6.4 Summary

It's easy to confuse square anything with anything squared, especially in casual speech, and this confusion is common in the news. Fortunately, in many cases the resulting value is so ridiculously large or small that you can see the error by thinking about consequences.

Be careful about how area grows when linear dimensions like height and width are changed. The rule is that area grows in proportion to the square of linear dimensions, so doubling the radius or the diagonal or both height and width will quadruple the area; increasing the radius by a factor of 10 will increase the area by a factor of 10 squared, or 100.

The factors are even bigger for volume and thus weight. Volume is proportional to the cube of the linear dimensions: doubling the radius of a sphere or all the sides of a box will grow the volume by a factor of 2 cubed, or 8, and if the linear dimension goes up by a factor of 10, the volume will go up by a factor of 1,000.

In all such cases, the important thing is proportion—it doesn't matter at all if there is some constant factor like 4/3 or π because it cancels out when one value is divided by the other. Nor does it matter what the shape is, since the difference between say a rectangle or triangle and a circle is also just a constant factor.

Chapter 7

Milestones

"Every day 10,000 baby-boomers turn 65."
(*New York Times*, August 1, 2014)

"8,000 baby-boomers turn 65 every month."
(*New York Times*, May 7, 2016)

Every day, thousands of newspapers print stories along the lines of "Every [some time period] [some number of] [some group] [does something]." Many of these stories are about a "milestone" or once-in-a-lifetime event, like birth, death, or a significant birthday.

7.1 Little's Law

Do you have any intuition about how many baby-boomers turn 65 every month? I don't either, but fortunately we can often reason about these formulaic statements and in this case we can even determine which one of the two quotations above is basically correct and which one is certainly wrong.

One technique is based on a rule of thumb called *Little's Law*, a conservation law that relates the number of things undergoing some process, the rate at which they arrive to be processed, and how long the process takes.

As a simple example that you can easily remember and use to check your understanding, imagine a school with 1,000 students. Each student enters, spends four years, then graduates. If we ignore dropouts and transfers, there are 250 students in each cohort, as you can see in the figure.

1,000 students in total

Figure 7.1: Little's Law for a 4-year school with 1,000 students.

Little's Law relates these three numbers: 1,000 students in all equals 250 students per year times four years. If we divide 1,000 by 250, we get four; if we divide 1,000 by four, we get 250; if we multiply 250 by four, we get 1,000. This relationship seems pretty obvious in retrospect, at least from this example, but apparently it was first described in 1954 by John Little, a professor in the Sloan School at MIT.

Let's use Little's Law to estimate how many American baby-boomers turn 65 every day. To simplify the arithmetic, we'll assume that the population of the US is 300 million; that's the number of people "in process." (The "process" is going through life.) Suppose that each person lives to age 75;

that's the processing time. These are over-simplifications, since some people die younger while others live longer, and it also ignores immigration, emigration and the birth rate, but it's good enough for now.

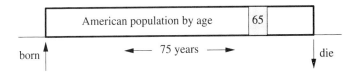

Figure 7.2: Little's Law for Americans.

If we divide 300 million by 75, we get 4 million people in each age group, which is the arrival rate (4 million people are born each year) and the departure rate (4 million people die each year). And 4 million is also the number of people who reach any particular milestone in a year, including turning 65. There are 4 million people who are any specific age like 65, as shown in Figure 7.2.

Dividing 4 million by 400 days in a year gives 10,000 per day. But there are only 365 days in a year (10 percent less than 400) so we can scale 10,000 up by 10 percent to conclude that about 11,000 people reach any given milestone per day.

Thus the statement that 10,000 baby-boomers turn 65 each day is reasonable, but the statement that 8,000 turn 65 every month is wrong; the units should have been days.

We can apply the same kind of reasoning elsewhere too:

"Every day this year, around 1,800 people will mark the iconic birthday [age 65], which for many heralds their retirement."

(*Daily Mail*, August 2, 2011)

The population of the United Kingdom is about 65 million, so we can simplify the arithmetic by assuming a lifespan of 65; that implies that one million Britons turn 65 every year, and thus about 2,700 turn 65 each day. But life expectancy in the UK is slightly over 80 years, so more like 2,300 people (65 million divided by 80 divided by 365) turn 65 each day. That value is higher than the 1,800 in the story, but not absurdly so; it's within a reasonable range.

For confirmation from independent sources, a number of stories about the birth of Prince George on July 22, 2013, observed that he was one of 2,200 babies born that day, though as a likely future king of England, somewhat unusual.

Note that we approached these computations by using approximate numbers for easy arithmetic, leaving open the option to refine them later if necessary. For example, I'm not sure about what part of the UK this was meant to include. Suppose that the relevant population was really 75 million. Then we could start with a life expectancy of 75 years (making the arithmetic easy), then adjust up or down as we learn more about the actual population.

This approach is really helpful; you should always look for ways to simplify the numbers up front, saving any necessary refinement for later on.

7.2 Consistency

Confirmation by independent computations or sources is good. The two statements at the beginning of this chapter were inconsistent, and that's a warning sign. There's not much difference between 8,000 and 10,000 as numbers but when units are attached, 8,000 per day is about 240,000 per month, so they differ by a factor of 25, which is a big difference, and thus a big clue that at least one of them is wrong.

By the same token, if a variety of independent computations produce consistent answers, that's a positive sign. Consider these examples:

"Every day, 10,000 baby-boomers turn 50 years old."
  (*Gambling Magazine*, May 1, 2005)

"Every week for the next 18 years, some 88,500 baby-boomers will turn 59½."
  (*Newsweek*, September 12, 2005)

"350,000 Americans turn 50 every month."
  (*Forbes*, January 10, 2005)

"4 million students graduate from high school each year."
  (*New York Times*, July 9, 2010)

These four values, expressed as per day, week, month, and year, are all within 10 or 20 percent of the estimated 11,000 per day, so there's a good chance that they are all correct.

As another example of how consistency can be used, let's look at some numbers on identity theft, which is a serious and perhaps growing problem.

"Every 79 seconds, someone becomes a victim of identity theft."
  (*CBS News*, January, 2001)

"Every two seconds, another American becomes a victim of identity fraud."
  (*CNN News*, February, 2014)

"19 people become victims of identity theft every minute."

(Security services company, 2015)

The third statement speaks of victims per minute, while the other two are about victims per second. These can't be compared directly, so our initial step should be to convert the third statement into the same units as the first two: 19 victims per minute is one victim every three seconds.

Every two or three seconds are roughly consistent rates, given that one quotation says "fraud" and the other says "theft," but both are much shorter than 79 seconds. Why might this be? Simple error is always an option, but another possible explanation is that things have indeed gotten worse over the course of 13 or 14 years. The "79 seconds" number, which is widely quoted, dates from 2001, a time that is comparatively early days in e-commerce; the other two numbers are much more recent.

It appears that the value 79 originally came from dividing the precise number of seconds in a year (31,536,000) by a round number from the US Federal Trade Commission—400,000 cases of some kind of identity theft. The result is 78.84, which was subsequently rounded up to the still overly-precise value 79.

In 2017, the FTC reported that it had received nearly 500,000 identity theft complaints, and the Department of Justice said that 17.6 million people were victims of identity theft. These two presumably authoritative numbers are sufficient to explain the apparent discrepancies. The FTC receives a complaint every 63 seconds; the DoJ says that someone is a victim every 1.8 seconds. Any discrepancy is likely because different things are being counted.

7.3 Another example

Not all "milestone" stories are wrong, of course, though sometimes it requires a bit of thinking to be sure. For example, in its July 2014 issue, the product-testing magazine *Consumer Reports* said that every day 130,000 Americans move into a new home.

My initial reaction was skeptical—that's a lot of relocation and the value must be too high. Fortunately, we can use the techniques above to come to something more objective than a gut reaction.

Suppose that each American moves exactly once in a lifetime. We know from our discussion of birthdays at the beginning of this chapter that this would mean about 11,000 people move each day. But personal experience tells us that most people move more than once in a lifetime; indeed, it would be quite unusual to have never moved. How often do people move? That varies a great deal, of course, but we can come up with a sensible range from our own experience. If we assume that a person moves say once every 6 or 7 years, that's 10 to 12 times in a lifetime; scaling 11,000 up by a factor of 12 comes close to 130,000 moves each day. The *Consumer Reports* value is likely to be sound.

7.4 Summary

Little's Law is an example of a conservation law: what goes in must come out, and if arrival rate, processing time, and number in process are all constant, there's a particularly simple relationship among them. Even if the assumptions are not perfectly true, as with populations of schools and countries, the approximation is good enough for reasoning about whether some statement is true or false.

Consistency of independent estimates, results, or anything else is a strong signal that the values are likely to be correct, unless there is some systematic error. Truly independent computations are not likely to involve a systematic error, so if you compute something in two different ways and you get much the same answer, that's a good sign. Something as simple as adding up a column of numbers once from the top and once from the bottom can help. To sum all the values in a table of numbers, add up the rows to get row totals and add up the columns to get column totals; the sum of row totals has to match the sum of column totals.

One way to do independent computations is to scale a big number down to individual items or people; in the other direction, compute from the bottom up, from how something affects a single item or person, up to all the items. For instance, from a story on public transportation in New York,

"The number of trips taken on public transport systems in 2008 was 10.59 billion, not million."
(*New York Times*, March 11, 2014)

The original value of 10.59 million is clearly wrong. Reasoning backwards, it would imply that each New Yorker took public transportation only once a year. You don't have to live in the Big Apple to know that's an error.

But would 10.59 billion trips be right? From the top down, if there are 10 billion trips in a year and 10 million New Yorkers, that's about 1,000 trips per person per year, or about three trips a day. From the bottom up, if a New Yorker makes two trips a day on public transport, multiplying by 365 gives about 700 trips a year, and multiplying by 10 million people gives 7 billion rides. That's not 10 billion, but it would be if the number were three trips a day. The 10.59 value seems reasonable. (I don't know whether New Yorkers think that a "trip" means a single ride or the round trip from home to work and back; that's a factor of two that might matter at some point but not here.)

Watch out for numeric impossibilities. According to the October 2008 edition of *Backpacker Magazine*,

> "14% of search and rescue incidents occurred on Saturday, the busiest day for search and rescue teams in the parks. 7% occur on Wednesday, making it the best day to have an accident."

Is this pair of numbers plausible or not? If 14 percent of incidents occur on Saturdays and 7 percent on Wednesdays, the remaining 79 percent must be spread across the remaining five days; that's a minimum of about 16 percent for each of those days. Thus at least one of them must be busier than Saturday. There's something wrong with the original statement. Saturday can't simultaneously be the busiest day yet less busy than the average of the five other days.

Chapter 8

Specious Precision

> "Hulu users had streamed 700 million hours of content in the
> first 90 days of the year. Divided into individual days,
> that's an average of 7,777,777.78 hours per day."
>
> (Blog post, August 2016)

> "He reached the summit of all 82 of the
> 13,123-plus foot peaks in the Alps in 62 days."
>
> (Multiple stories about a climber, May 2017)

Something about the numbers in these two excerpts catches the eye: some of them are very precise. That makes them fine examples of a particular kind of innumeracy called *specious precision*: numbers that are presented with more precision than they really have.

The Oxford Dictionaries web site defines *specious* as "Superficially plausible, but actually wrong. Misleading in appearance, especially misleadingly attractive." (Another dictionary site says that very few English speakers will know this

word. I don't believe that, but just in case, now you have a new
word in your vocabulary.)

Speciously precise numbers are usually a sign of some combination of ignorance and laziness, though from time to time
there's also an attempt to mislead. Let's look at some examples.

8.1 Watch out for calculators

Consider the number of hours per day of Hulu streaming.
Hulu is a video-on-demand service with over 10 million subscribers. If Hulu streamed 700 million hours in 100 days, that
would be 7 million hours in one day; if the period was 90 days,
then the value would be roughly 10 percent higher because we
would be dividing by 90 instead of 100. If I had to compute
hours per day, that's how I would do it.

But clearly this value was computed by using a calculator;
in fact, Figure 8.1 shows exactly how it came about, dividing
700 million by 90. (When I first tried this computation with
the calculator program on my Mac, the default setting for precision gave me a full 15 meaningless digits after the decimal
place: 7777777.77777777777778!)

The original numbers that went into this computation had at
most a single digit of precision: 700 million (more than 600
million, less than 800 million), and 90 days. (In non-leap years
there are exactly 90 days in the first three months, so "90"
could be exact or it could be an approximation to a quarter of a
year.) So the resulting ratio can't have much more than one
digit of precision either. There are many better ways to say
this: 7 million hours per day, or 8 million, or perhaps 7.7 or 7.8
million or "seven and a half million." Any of these would be

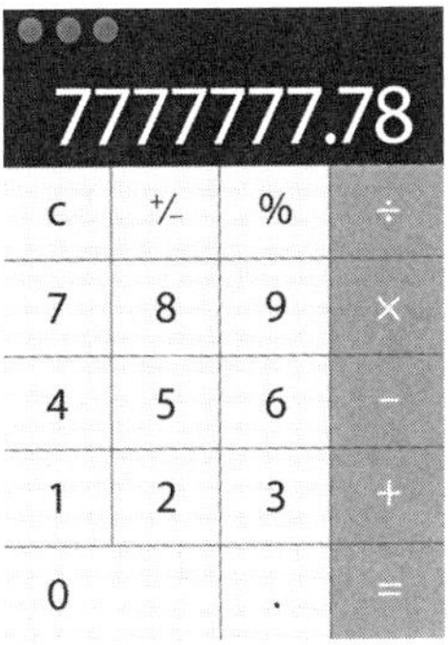

Figure 8.1: 700 million divided by 90.

defensible, but not 9 digits copied off the screen of a calculator or some web-based equivalent.

The basic rule is that a computational result should not be presented with more precision than the input values warrant. If your original data has only single-digit accuracy, don't expect much more than that for the result.

8.2 Units conversions

Let's look at the second example at the beginning of this chapter. There are three apparently precise numbers: 82 peaks, 62 days, and 13,123 feet. The first two are presumably accurate, especially if one is counting something discrete and reasonably well-defined like mountain peaks or days. But what's special about 13,123 feet? Why would mountain climbers be so interested in that particular value, while dismissing peaks that were only say 13,100 feet high?

The answer is that 13,123 feet is 4,000 meters, a nice round number. People like round numbers—they are easy to

Figure 8.2: More than 13,123 feet high?

remember and they convey the essence of a value without excess precision. But suppose that you are a reporter in the US doing a story on Alpine climbers. In Europe, heights are measured in meters, but the US still uses the English system of units, so someone has to convert the original round number for American readers. Out comes the calculator again, and sure enough, 4,000 meters is a little over 13,123 feet (Figure 8.3).

Problem solved, readers informed? Not really. A better approach would be to give both values, something like "4,000-meter (13,123-foot) peaks," thus conveying more information and perhaps educating readers along the way.

This particular kind of metric-English conversion is common in the US. For instance, a story in the *New York Times* in March 2008 quoted the editor of *The Yacht Report* (not a

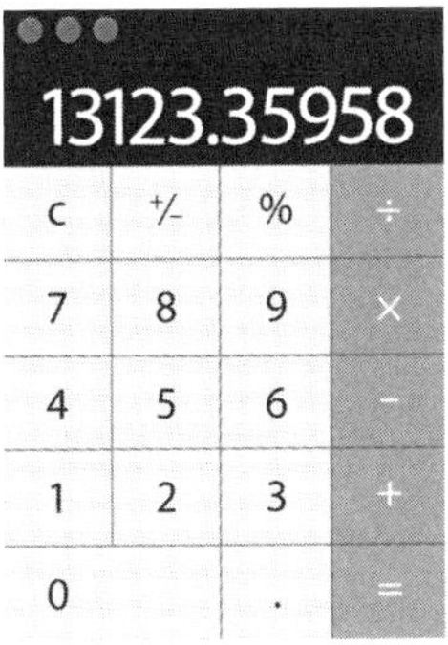

Figure 8.3: Converting 4,000 meters to feet.

publication that I read regularly) as saying "When a yacht is over 328 feet, it's so big that you lose the intimacy." We can imagine a yacht owner saying "I used to have a 300-footer, and it was wonderful—I got to know everyone really well when we sailed; it was like a big family gathering. But my new 328-footer is so big that I hardly know who's on board." Where did 328 come from? Of course it's 100 meters, a nice round number of the sort we use in everyday speech. If the story had originated in the US, it would probably have been "over 300 feet" or "over 100 yards."

Spotting multiples of 328 can easily become a kind of nerdy party game. For example, a story about FCC regulation of mobile phones in the US says "Carriers using handset-based technology must locate 67 percent of calls to within 164 feet. Carriers who use network-based technology are allowed an accuracy standard of 328 feet for 67 percent of the calls." A story on runway widths at an airport in India says "the runway is 656 feet wide; the Indian government standard is 984 feet." Do the conversions back to metric, and you'll see that all of

Figure 8.4: An intimate 68-meter (223-foot) yacht.

these numbers are multiples of 50 or 100 meters.

What other numeric conversion factors are you likely to see? In 2013, the technology news site *Slashdot* reported that "Ferrari has unveiled its fastest car ever, a nearly 1,000 hp. gas-electric hybrid dubbed LaFerrari that does 0-62 mph in less than 3 seconds, 0-124 mph in less than 7 seconds, 0-186 mph in 15 seconds." Where do these not-round numbers come from? Well, Ferraris are made in Italy, where speeds will be expressed in kilometers per hour, and a kilometer is 0.62 miles. Zero to 100 km/hour would be a standard kind of comparison, though I imagine that zero to 300 km/hour is more for bragging rights than for everyday driving. We've spotted another blind conversion from metric, and can add the factor 0.62 to our collection.

Of course conversions go in the other direction too. Just as Alpine peaks over 4,000 meters are interesting, there are analogous numbers in the US, for example, the 46 peaks of the Adirondack range that were historically reckoned to be over 4,000 feet high. Sure enough, it's easy to find stories from outside the US that talk about peaks that are "over 1,220 meters high." Or this from a 2009 article on search and rescue operations from the *Journal of Travel Medicine*: "the most common rescue environments were mountain areas between 1,524 and 4,572 m." Those are 5,000 and 15,000 feet.

From an article about military weapons, "The M16A1 does not put enough spin on the heavier M855 bullet to stabilize it in flight, causing erratic performance and inaccuracy for training or full combat usage. It should only be used in a combat emergency and then only for close ranges of 91.4 meters or less." If I were under fire, I would certainly prefer a round number, perhaps the length of a football field, to having to think about "91.4 meters."

Finally, Figure 8.5 shows a metric-conversion picture that I took in a local toy store.

If blind calculation is over-used for converting between metric and English units of length, it's equally abused for metric and English weights. In April 2016, the *Daily Mail* reported that "Apple recovered 2,204 pounds of gold from recycled electronic devices last year...worth a cool $40 million."

Forty million is a nice round number, but now that you've been sensitized, does 2,204 seem unusually precise? And indeed it is. One kilogram is 2.204 pounds, so multiples of 2.2 and 2.204 can be spotted whenever there's a conversion from metric weights to English. The original value was surely 1,000

Figure 8.5: Speciously precise English to metric conversion.

kilograms and the price of gold must have been about $40,000 per kilogram.

The story goes on to say that "Apple also recovered 6,612 pounds of silver, 2,953,360 pounds of copper, and 23,101,000 pounds of steel." The first two values are multiples of 2,204 as well; the third value has gone through some other process on its way to excess precision.

Drug busts are a popular source of weight conversions, as in these headlines from 2017:

"Man found with 22 pounds of pot gets probation"

"2 arrested after 22 pounds of cocaine found at home"

"44 pounds of meth found during traffic stop"

"55 pounds of cocaine, worth $750,000, found after traffic stop"

Values like 22 and 44 pounds were almost surely originally 10 or 20 kilos. The final one, 55 pounds, is presumably a nice round 25 kilos, which at a nice round $30,000 per kilo has a nice round street value of $750,000.

English-metric drug conversions go in the other direction as well; a story in May 2017 describes how the US Coast Guard seized "an estimated 454 kilograms of cocaine," a value that undoubtedly started life as 1,000 pounds.

Figure 8.6: *Rhymes with Orange* Copyright © 2008, Hilary B. Price. Distributed by King Features Syndicate.

8.3 Temperature conversions

Our examples so far have all involved simple multiplicative factors: 2.2 for weight, 3.28 for length, and so on. By contrast, temperature conversions between Celsius and Fahrenheit are a bit more complicated because zero for Celsius is not zero for Fahrenheit. That leads to a different kind of confusion, exemplified by this comment from a climate change web site.

"If 1 degree Celsius equals 33.8 degrees Fahrenheit, so wouldn't .5 degrees Celsius equal about 17 degrees F? Since 1 degree Celsius equals 33.8 degrees F, and the chart shows the upward trend beginning around 1980, if the average temperature rise during that period is only a half degree Celsius that would be about a 17 degree rise in Fahrenheit. Move your air conditioning thermostat up 17 degrees and feel the difference."

The implication is that a climate change of half a degree in Celsius would be an enormous 17 degree change in Fahrenheit, which would be very noticeable, and thus climate change can't be real (or it would be really obvious—I'm not sure which point the writer intended).

This excerpt from a book on mountaineering has exactly the same problem: "For every 100 meters (330 feet) one gains in altitude, the temperature drops by about 1 degree centigrade (33.8 degrees Fahrenheit)."

The confusion here is linguistically reminiscent of the "square feet, feet square" problem that we talked about in Chapter 6; it's the difference between 1 degree Celsius (a specific temperature) and 1 Celsius degree (a difference between two temperatures).

When the temperature changes by 1 Celsius degree, it changes by 1.8 Fahrenheit degrees. Thus if the temperature goes from 1 degree Celsius to 2 degrees Celsius, it goes from 33.8F to 35.6F. And of course it's exactly the same idea in the other direction.

If the conversion is not a simple multiplication, pay attention; it's more complicated to get things right.

8.4 Ranking schemes

"Princeton University is ranked #1 in National Universities. Schools are ranked according to their performance across a set of widely accepted indicators of excellence."

(*US News* 2017 rankings of US national universities)

It's always nice to have one's merits appreciated and publicly acknowledged. Princeton has been ranked first among national universities every year since I first started teaching there in 1999, except for one year when I was on sabbatical, and another year when *US News* appears to have made a clerical error that dropped it into second place.

Of course it's ridiculous to say that Princeton is the best university in the country. It's a fine school with much to offer its students, but it's only one of many fine schools, and the characteristics that make it particularly good for one student might be less important for another.

University rankings are one example of a common bit of too-precise computation. Another popular example is ranking places to live. A search for "best cities to live in" produces myriad articles about cities that are deemed especially desirable as places to live and work. But surprisingly, or perhaps not, there's little agreement about the results. There was literally no overlap among the top five cities on the first half dozen lists that my search produced.

The process by which such rankings are produced is straightforward in outline. Decide on factors that are thought to be important—tuition, standardized test scores, class sizes and endowment for colleges; or housing prices, school quality, public transportation, and cultural amenities for cities. Collect data on each factor and convert it into numeric values. Assign weights to each factor—perhaps test results should be worth 25% of a school's scores while endowment size might be 10%. Combine the individual factors with the weights to compute a single number for each school or city, and sort them into decreasing order. The name at the top of the list is the best place to go to school or to live.

This process makes it clear why there's so little agreement about schools or where to live. We've collected data that is often flaky (how do we measure housing prices or teaching quality?), converted non-numeric data into numbers (cultural amenities as a number?), and combined them with arbitrary weights (why not 20% and 15% instead of 25% and 10%?). Flaky data plus arbitrary weights is a recipe for shaky results.

One of the reasons why Princeton consistently ranks high is that a significant component of the *US News* ranking is based on the rate at which alumni give money. Princeton alumni are remarkably loyal and generous—about two thirds of them give money every year—so if alumni loyalty were the only factor, Princeton would always rank first in this group.

It's not that there isn't a grain of truth in rankings, of course, but it's silly to put much credence in specific placements, and certainly no reason to believe that adjacent values are in the right order.

The *Places Rated Almanac* was an annual publication that attempted to rank 329 US metropolitan areas on their desirability as places to live, based on nine factors like climate, housing costs, crime rate, and public transportation. In 1987, four statisticians from Bell Labs published a paper called "Analysis of Data from the Places Rated Almanac." The authors showed that by suitably adjusting the weights of these factors, it was possible to put any one of 134 cities in first place, and any one of 150 cities in last place. Remarkably, 59 cities could be ranked either first or last by suitable choices of weights. Since then, whenever I see a ranking based on multiple weighted factors, I say to myself "places rated," and treat the conclusions with considerable skepticism.

8.5 Summary

"There is no surer indicator of scientific illiteracy than the quotation of numerical data to a degree of precision greater than the experimental observations warrant."

(Peter Medawar, Nobel Prize-winning biologist)

When a number is expressed with high precision, that suggests that it is in some way more accurate than if it were written with lower precision, and thus that the number is more important or significant. It has subliminally acquired an unwarranted authority.

Precision and accuracy are not the same thing. Here's a conversation between a friend and Alexa on his Amazon Echo:

Friend: "Alexa, what is the snow forecast for today?"

Alexa: "Snow is very likely today. There is a 78% chance of snow, and you can expect about 0.73 inches."

Friend: "Wow, she's really accurate."

Alexa's numbers are certainly precise, but I'm sure that your experience with weather forecasting will tell you that she's not likely to be that accurate.

One harmless example of excess precision comes from the covers of magazines, which are fond of teasers like "43 ways

to pay less for practically anything" (*Consumer Reports*) or
"487 hot new looks" (*Harper's Bazaar*). Market research must
have shown that this kind of fake precision induces more people to buy the magazine than round numbers do.

Newspapers are not immune to attention-grabbing "precision" either:

$1,101,583,984.44

"That's the amount of unpaid securities fines in Canada,
a Globe investigation has determined. Regulators mete
out $100-million in new fines each year to generate
tough-on-crime headlines but collect no more than a
fraction."

   (*Toronto Globe and Mail*, December 22, 2017)

That's a remarkably precise figure, no less than 12 "significant" figures, and it must be important, since in the original
paper it was printed in letters half an inch high, definitely a
way to grab a reader's attention. The story goes on to describe
how regulators collect only a tiny fraction of fines that are due,
an amount that the paper computed by combing through 30
years of records.

The precision is a bit illusory; the story goes on to say
"However, The Globe was unable to obtain complete historical
data on unpaid fines from every regulator, so the true number is
likely even higher."

So if the true number is likely even higher, why 12 digits?
Presumably all those digits will grab more attention than a
mundane headline like "Over a billion dollars of unpaid securities fines."

Many examples of overly-precise numbers arise from mechanical conversion between different systems of units; others come from blind copying of the numbers off the display of a calculator, often without taking into account the precision (or lack thereof) of the original numbers. Both are bad practices.

If you combine approximate data, sometimes not even numerical, with arbitrary weighting factors, you can produce rankings that are good for starting spirited discussions, but nearly useless for drawing meaningful conclusions. Treat all ranking schemes with a grain of salt.

Figure 8.7: *Dilbert* © 2008 Scott Adams. Used by permission of Andrews McMeel Syndication. All rights reserved.

Chapter 9

Lies, Damned Lies,
and Statistics

"The average Yaleman, Class of '24, makes $25,111 a year."
(Darrell Huff, *How to Lie with Statistics*, 1954)

This is the first example in Darrell Huff's wonderful little book, a marvelous introduction to statistical chicanery that is just as instructive and fun to read today as it was when published over 60 years ago.

Huff's title is an allusion to the famous aphorism "There are three kinds of lies: lies, damned lies, and statistics." This is often attributed to Benjamin Disraeli, who was Prime Minister of England from 1874 to 1880, though its first documented appearance was not until 1891, well after his death in 1881.

No matter who said it first, it captures a justifiable cynicism about how statistics can be used, whether consciously or not, to mislead. In this chapter we'll look at some examples. This is definitely not a statistics book, but we'll look at a handful of basic statistical ideas that will help you defend yourself if you understand them.

9.1 Average versus median

Huff offered two complaints about the value $25,111. The first was that it was "surprisingly precise," which echoes the theme of the previous chapter. One imagines that someone surveyed a bunch of Yalies about their annual incomes, added up the resulting numbers, and divided by the number of respondents. That sounds rather like our discussion about the blind use of calculators, doesn't it?

I don't know my annual income, though I could make a guess. I'll have a more accurate number when I work on my taxes but it still won't be exact. You're probably in a similar situation.

But if you were going to fill out a survey for an alumni publication, would you provide the comparatively precise number that you sent to the tax authorities? Of course not. If you bothered to reply at all, you would make a rough guess and round it to only one or two significant digits. Adding up a bunch of rounded approximate values, then presenting the average to the nearest dollar or pound or euro, is a good example of specious precision.

There's another potentially serious problem—if there are a few real outliers in the group, their values can skew the average a lot. Suppose we want to compute the average net worth of Harvard dropouts of the past forty or so years. I will conjecture that people who dropped out of Harvard haven't done as well on average as those who got through the full four years, but there are a pair of notable exceptions: Bill Gates, founder of Microsoft, and Mark Zuckerberg, founder of Facebook, whose combined net worth is at least $150 billion.

How many Harvard dropouts are there? Harvard has 6,600 undergraduates, so about 1,650 students enter each year. (Remember Little's Law?) The six-year graduation rate for Harvard is around 97%, so only 3% of the students in any given cohort fall by the wayside; that's about 50 for each year, or 2,000 in forty years.

Suppose that the net worth of these comparative unfortunates is $100,000 each, or $200 million in total.

In the spirit of the original Yale story, then, the total net worth of the Harvard dropouts is $150,200,000,000; dividing that by 2,002 gives an average of $75,024,975. This figure might be technically correct but it's terribly misleading, as averages tend to be when there are a handful of far-out values.

There's a better way to characterize a set of numbers like this: the *median*, which is the value at the middle of the group. There are as many values less than the median as there are greater than the median. The median of our hypothetical population of dropouts is $100,000; the presence of Mr Gates and Mr Zuckerberg has no effect on that, and indeed the median would not shift if we added another few hundred truly wealthy people or a bunch of paupers.

When you see *average* (or its more formal synonym, *mean*), be cautious; outliers may be skewing the result. The ordinary arithmetic average works well when the values are nicely distributed, like heights and weights for large groups of people. But it's not as good if there are significant outliers. For those, the median is a more representative statistical value. Half the values are below the median; half are above.

Figure 9.1: Average Harvard dropouts?

"The median grade at Harvard College is an A−, and the most frequently awarded mark is an A."
(*Harvard Crimson*, December 3, 2013)

Another characteristic number is the *mode*; it's the value that occurs most frequently. At Harvard, the mode is an A.

9.2 Sample bias

"According to AARP magazine, some 48.7 percent of people age 55 and over who were surveyed said they like to participate in surveys."
(*New York Times*, November 12, 2005)

Huff had another observation about the survey of Yale graduates of the class of 1924; he described the reported average income as "quite improbably salubrious." From today's perspective, $25,000 sounds like minimum wage, but if we convert from 1954 dollars to 2018 dollars (using usinflation.org), it would be more like $230,000.

Huff's conjecture was that survey respondents would disproportionately be the successful ones; alumni who had not prospered would be reluctant to share their lack of success with their classmates and might also have been harder to track down. So the average would likely be based on a biased sample: those who had been comparatively successful.

Similar caveats should apply to the AARP results as well. If fewer than half of the people who participated in this survey liked to participate in surveys, that suggests that more than half didn't like to do surveys, and it's safe to assume that there were many others who declined to participate at all. Over the population at large, the number of people who like to participate in surveys is likely to be even smaller. (We don't know how many people were surveyed either. The smaller the sample size, the less likely it is to be meaningful.)

Sample bias or sampling error is at the heart of many prediction failures. One of the most famous was in the US presidential election of 1936. The magazine *The Literary Digest* predicted that the Republican candidate, Alf Landon, would win the election by a large majority, based on a survey sent to 10 million subscribers, with 2.3 million responses. As it turned out, however, the Democratic candidate Franklin Roosevelt won by the largest majority in modern times.

Statisticians and politics junkies have studied this polling failure ever since. Among the factors that contributed to *The*

Literary Digest's mis-prediction, it appears that its readers were disproportionately Republican and also more interested in politics than the general population was. Thus the sample was biased towards Republicans to begin with, and those who responded were predominantly strongly anti-Roosevelt. The sample, though very large, was far from representative.

By contrast, George Gallup began his national polling career with a successful prediction for this election, based on a better-chosen sample of only 50,000 potential voters. *The Literary Digest* folded in 1938; the Gallup Poll continues to this day.

For the 2016 US presidential election, polling was far more sophisticated than in the 1930s, and the consensus was that with high probability Hillary Clinton would be elected. As it turned out, although Clinton won the popular vote by about 3 million votes, Donald Trump won the Electoral College and thus the presidency. Did pollsters miss some important constituency that went for Trump, or were people not being honest in their responses to pollsters, or did voters change their minds at the last instant? Statisticians and politics junkies will also be studying this election for years to come.

9.3 Survivor bias

"The message that 'Smoking Kills' is just a lie. I've been smoking for 45 years, and it still hasn't killed me. And in fact, throughout those 45 years, I've never had any serious diseases at all. No cancer. No heart disease. No emphysema. No dementia. No arthritis. No nothing."

(Wordpress blog, 2016)

"Cigarette smoking is the leading cause of preventable disease and death in the United States, accounting for more than 480,000 deaths every year, or 1 of every 5 deaths."

(Centers for Disease Control and Prevention, 2017)

Can you get rich by working very hard? Bill Gates and Mark Zuckerberg did. Can you pick stocks far better than most people? Legendary investors like Warren Buffett invest successfully for decades. Can you live to a ripe old age by heavy smoking and drinking? Some centenarians say so.

We can't safely generalize from cases like these, however, because they are examples of *survivor bias*—the data that we are generalizing from is not representative, but rather cherry-picks from instances that have survived. Data that would lead to a different and more accurate conclusion has been removed from the population since it didn't survive. Our apparently healthy blogger is a lucky survivor, not proof that smoking is harmless.

9.4 Correlation and causation

One of the most frequent statistical errors is to get cause and effect wrong. Just because two things seem to vary in proportion to each other does not mean that one causes the other. There's a very funny web site, tylervigen.com/spurious-correlations, that displays a large number of correlations that are not based on any causal relationship. For example, from 2000 to 2009, the divorce rate in Maine correlated almost perfectly with per capita consumption of margarine, and the money spent on pets in the US correlated almost perfectly with the number of lawyers in California.

Those are both pretty clearly nonsense, but what about this?

"Soda boosts violence among teens, study finds."
(*Washington Post*, October 23, 2011)

The article goes on to say that "heavy use of carbonated non-diet soft drinks was significantly associated with carrying a gun or knife and violence towards peers, family and partners." The study was based on a small sample (1,800 students in Boston), which already makes it a bit dubious. But the real issue is that there are myriad other factors, like low socio-economic status, that could readily explain both a tendency to violence and poor dietary choices. The study found some correlation ("significantly associated with"), which the headline converted to a false conclusion, that soda "boosts violence." This kind of leap from correlation to causation is not uncommon in the news; you have to be alert for it.

The core principle is that *correlation does not imply causation*. A strong correlation between smoking and increased cancer risk had been observed for many years, but it took some while before the mechanisms of cell damage became well enough understood to explain how smoking could trigger cancer. It seems likely that we are seeing the same kind of process with climate change, excess sugar in our diets, and a number of other issues.

9.5 Summary

Statistics is a large field and its proper use requires training and experience. The handful of topics in this chapter are the bare minimum for defending yourself against invalid statistical claims and reasoning.

The arithmetic average or mean is fine for characterizing a set of numbers, but the median is sometimes better, since it is the value at the middle of the set and thus less subject to the effects of extreme outliers like Messrs Gates and Zuckerberg.

Most statistical results are based on samples of a population, not the entire group. That makes them vulnerable to potentially serious sampling errors unless the samples are truly representative. Pollsters know this, of course, but it's still easy to fall short and draw a conclusion from a sample that isn't true of the whole population.

Survivor bias is another form of sampling error. Excluding items, whether consciously or unconsciously, because they are not thought to be relevant or just aren't currently in the population, can lead to misleading results, often overly optimistic.

Correlation does not imply causation. If two things appear to change in step with each other, that does not mean that one causes the other. There could be a third factor that affects both, or it could be mere coincidence, as with divorce and margarine.

Figure 9.2: Correlation. Copyright © 2009, Randall Munroe, *xkcd*.
Source: http://xkcd.com/522.

Chapter 10

Graphical Trickery

"[A misleading graph] is vastly more effective, however, because it contains no adjectives or adverbs to spoil the illusion of objectivity. There's nothing anyone can pin on you."

(Darrell Huff, *How to Lie with Statistics*, 1954)

How to Lie with Statistics illustrates several graphical techniques that are widely used to mislead or deceive. Today, with computers, chart-drawing tools like Excel, and picture manipulators like Photoshop, deception can be far more sophisticated than would have been possible 60 years ago, and I imagine that Huff would find plenty of new material for an updated version of the classic.

In this chapter, we're going to look at a handful of such tricks, most of which Huff mentioned. Once you see a few examples, you'll start to see similar ones all over the place, and you'll be better equipped to defend yourself. And as with some of the other things we've talked about, it's fun to discover examples in the wild, so you might find that you read the news and surf the web with a more critical eye.

10.1 Gee-whiz graphs

On May 6, 2010, stock markets in the United States suffered a dramatic fall in stock prices, a terrifying plunge now called a "flash crash." In the graph below, the vertical axis is the Dow Jones stock price index and the horizontal axis is time of day.

As you can see, stock prices fell almost to nothing at around 2:45 pm, before recovering to about three quarters of their previous value by the time the market closed at 4 pm. The bottom just dropped out of the market for a short while.

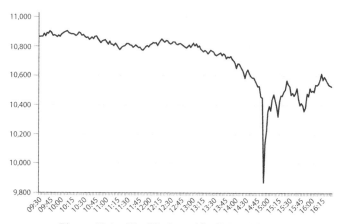

Figure 10.1: The "flash crash" of May 5, 2010.

If you look at this graph more carefully, however, you'll see that the vertical axis doesn't start at zero; the origin is at 9,800. Starting there instead of at zero greatly exaggerates the vertical scale and gives an impression of far more change than there really was; that's why Huff called them "gee-whiz" graphs.

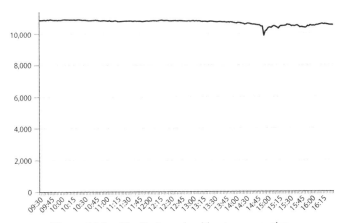

Figure 10.2: The flash crash without exaggeration.

If we restore the zero value to the vertical axis, we get a rather less dramatic picture of the drop, in Figure 10.2. It would still be scary for investors, but the new graph shows that we're talking about a maximum drop of under 10 percent and a net of just over 3 percent, not the end of the world.

Gee-whiz graphs are not always so dramatic. For example, Figure 10.3 shows the growth of the number of millions of Twitter users active each month, redrawn from the Form S-1 that Twitter filed with the US Securities and Exchange Commission as part of the process of going public in 2013.

It looks like the number of users has tripled over 18 months, since the rightmost vertical bar is about three times as high as the leftmost. Setting the origin to zero instead of 100 gives a different impression, however: the growth is from 138 to 215 million, which is a factor of 1.56 (Figure 10.4).

Edward Tufte, author of *The Visual Display of Quantitative Information* (1992), has a different opinion about whether the

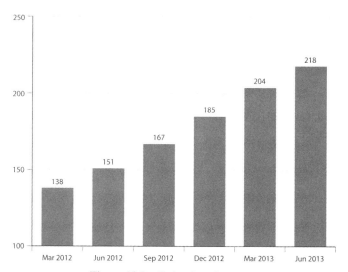

Figure 10.3: Twitter's active users.

zero value should be shown:

> "In general, in a time-series, use a baseline that shows
> the data not the zero point. If the zero point reasonably
> occurs in plotting the data, fine. But don't spend a lot
> of empty vertical space trying to reach down to the zero
> point at the cost of hiding what is going on in the data
> line itself. (The book, How to Lie With Statistics, is
> wrong on this point.)"
>
> (www.edwardtufte.com/bboard)

There's no single right answer, but you should be aware that
a gee-whiz graph makes small differences appear large, and
that's often deceptive. It may well be done with the best of
motives—after all, a large expanse of completely empty graph

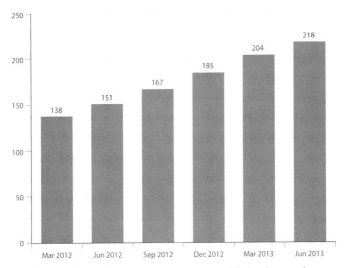

Figure 10.4: Not so gee-whiz version of Twitter's growth.

area isn't likely to draw the reader's eye, and an expanded scale makes it easier to see fine detail—but to me it still seems a bit questionable.

10.2 Broken axes

Sometimes you'll see a graph like the ones above but with a vertical axis that shows zero at the bottom, then a little jagged line to indicate that the axis has some missing ticks, and then back to the gee-whiz part. The break in the data is very clearly marked in the example of Figure 10.5; in other instances, the issue is more subtle.

Less commonly, you might see a graph where the horizontal axis is broken in the same way as the vertical axis, or you may

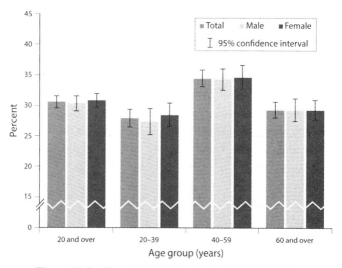

Figure 10.5: From National Center for Health Statistics.

see a non-uniform set of horizontal ticks; that's especially pernicious, since it tends to make the graph smoother than it really is and thus implies a more uniform process than exists. We'll see a nice example in Chapter 11.

10.3 Pie charts

Pie charts are often used to show how something is divided among a set of disjoint choices—the area of each slice of the pie corresponds to its fraction of the pie—but a pie chart can also be an effective way to obscure or misrepresent. The most common problem occurs when the chart is displayed in a perspective view, because this distorts the areas: sections towards the front look bigger.

Consider the two charts in Figure 10.6. They represent exactly the same data: four equal values, so each slice is 25 percent of the whole. The one on the left shows that accurately, since the area of each sector is one quarter of the pie. The chart on the right distorts the data: the visual appearance of the bottom two sectors is much larger than the top two.

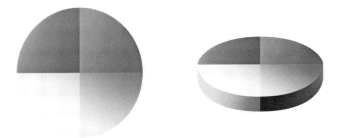

Figure 10.6: Two pie charts with the same data.

Of course the values in the pie chart should add up to 100 percent. It's not clear what to make of the example in Figure 10.7, derived from a *Fox News* visual, where support for the three candidates adds up to 193 percent.

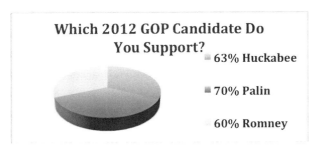

Figure 10.7: Really broad support?

10.4 One-dimensional pictures

In the US, the federal government subsidizes students from low-income families with financial aid that helps them to attend college, through a program called Pell Grants. The picture in Figure 10.8, from a Princeton University press release in 2016, shows that the number of incoming freshmen who are eligible for Pell Grants increased dramatically from the class of 2008 to the class of 2020.

Figure 10.8: Pell Grant eligibility increases enormously!

Or did it? If we ignore the visuals and focus on the actual numbers, we can see that Pell eligibility of the incoming class increased from 7 percent to 21 percent over a 12-year period. That factor of three really is commendable, but the visual impact of the two circles exaggerates the improvement. As we saw in Chapter 6, area grows as the square of the radius, so the circle on the right has nine times the area of the one on the left, and the text is far bigger as well. A casual reader is likely to take away the impression that there was an order of magnitude improvement, not the more accurate "factor of three."

This is an example of what Huff called a *one-dimensional picture*. Data values are represented by a graphical presentation that uses area or even volume to display values that should be shown on a linear scale. One-dimensional pictures are often used to suggest more impact than is warranted, though they are sometimes merely a misguided attempt to make mundane numbers look interesting. Compare Figure 10.8 to the minimal graph in Figure 10.9, which accurately represents the two values by their heights:

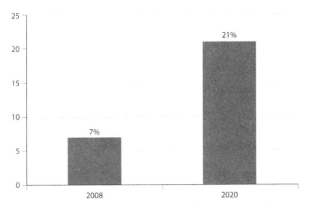

Figure 10.9: Percentage of Pell-eligible freshmen triples.

Boring, isn't it? But it conveys the information without misleading anyone. If you're fond of circles, however, make the *areas* proportional to the values, and use the same font size, as in Figure 10.10.

The area of the circle on the right is three times the area of the one on the left, which is an accurate representation of the corresponding numeric values.

Of course if there are only two values, a graphic contributes next to nothing; it would be sufficient to say that Pell Grant eligibility rose by a factor of three, from 7 percent for the class of 2008 to 21 percent for 2020.

Figure 10.10:   Percentage of Pell-eligible freshmen triples.

Using area to misrepresent a linear measurement is bad
enough, but it's quite possible to do worse. Figure 10.11
shows one of my all-time favorites. Summer stipends for graduate students went up by a factor of almost four, from
$500,000 to $2,000,000, as measured by the height of each
tiger (Princeton's mascot).

That's the total content of the image: two numbers, one of
which is four times bigger than the other. The picture makes
the increase look enormous, however, because the linear values
are represented by 3-dimensional tigers, and our eyes see not
the factor of four, but the increase in volume, which is 4^3: the
tiger on the right has 64 times the weight of the one on the left.

10.5 Summary

A picture is worth a thousand words, so presumably a misleading picture is worth a thousand misleading words. We've
seen how numeric data can be presented graphically in ways
that give the wrong impression. Our examples are far from all-

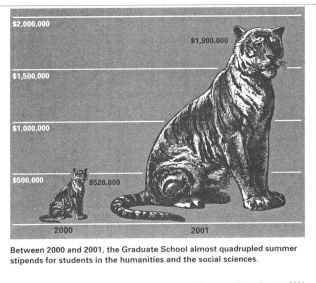

$2,000,000

$1,900,000

$1,500,000

$1,000,000

$500,000 $526,000

2000   2001

Between 2000 and 2001, the Graduate School almost quadrupled summer stipends for students in the humanities and the social sciences.

GRADUATE NEWS, SUMMER 2001

Figure 10.11: Huge increase in summer stipends!

inclusive: modern technology has made it easy to produce a rich variety of attractive and appealing displays both good and bad.

For example, in the chart of Figure 10.12, the cones (three-dimensional objects) are not as deceptive as they might be; because they all have the same base, the volume is proportional to the height. But the weird offset of the vertical axis is really misleading. It would not be necessary to label the individual items if the labels on the vertical axis lined up with the horizontal grid lines.

What should you watch out for? The most common case is likely to be the gee-whiz graph, where the vertical axis runs

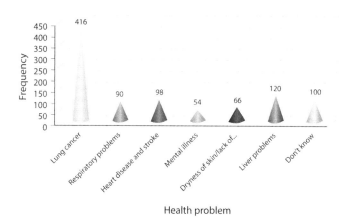

Figure 10.12: Actively misleading or just confusing?

only over the data values, largely eliminating the baseline value of zero. That has the effect of amplifying changes, generally making them seem more significant than they really are. It's at best a small improvement when the missing scale is hinted at by some indication that the vertical axis is broken.

You'll sometimes see something similar on the horizontal axis, where the values along the axis are not at uniform intervals. This trick is used to make trends look smooth and regular, where the real data is not.

Watch out for pie charts that are drawn with a perspective or 3-dimensional effect. That distorts the information, making the values at the front of the pie look bigger than those towards the back.

Watch out for one-dimensional graphs that use area or volume to represent linear values. Our eyes see area and volume, so it's easy to take away the wrong impression.

Even when there's nothing obviously suspect, be cautious. Look carefully at the graph in Figure 10.13:

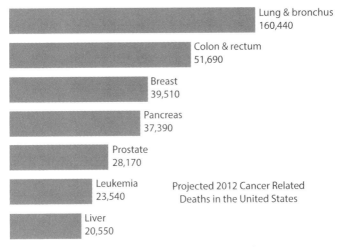

Lung & bronchus
160,440

Colon & rectum
51,690

Breast
39,510

Pancreas
37,390

Prostate
28,170

Leukemia
23,540

Projected 2012 Cancer Related
Deaths in the United States

Liver
20,550

Figure 10.13: http://www.medicalnewstoday.com.

There's no horizontal scale, but if the numbers are right, the top bar should be more than three times as long as the second bar. The other bars appear to be in the correct proportion to each other, so why was the top one shortened arbitrarily? Perhaps it was done for esthetic reasons, to avoid having one bar dominate the whole display. But that robs the graph of its most important message: the first category, lung cancer deaths, is bigger than the next four combined.

There's no shortage of graphical trickery, whether conscious attempts to mislead or merely misguided attempts to be visually appealing. But once you've seen and recognized some good representative examples, you'll be armed for the future.

Chapter 11

Bias

"Four thousand teens will try their first cigarette today."
(Advertisement, *New York Times*, November 18, 2005)

"Every day 5,000 teenagers try pot for the first time."
(Advertisement, *New York Times*, November 4, 2005)

Those two full-page advertisements caught my eye, in part because they came only two weeks apart, and each occupied the last page of one section of the *Times*. They were hard to miss.

Can we assess their likely accuracy? The first step is to apply Little's Law, since they are both in the "every time period something happens" category. Suppose that if a teenager is going to try smoking, it will happen on his or her 13th birthday. (This was approximately true in my case; fortunately my mother caught me and read me the riot act, for which I am eternally grateful.)

How many kids turn 13 every day? We've seen that in the US that's about 11,000, so to simplify the arithmetic, call it 12,000. If a third of teens try smoking, that's 4,000. The estimate seems to be reasonable or perhaps somewhat high. The CDC says that in 2016 about 15 percent of the adult population smokes, and that fraction has been decreasing, so a similar advertisement today might have a lower number.

11.1 Who says so?

These full-page advertisements cost someone a substantial amount of money. Who paid for them? I don't know for sure, but the one about cigarettes said "Endorsed by American Academy of Pediatrics, American Heart Association, American Lung Association, American Medical Association, National PTA." That's a heavyweight group of supporters whose concern with a major public health issue comes from a variety of perspectives.

The other advertisement, about the 5,000 teenagers who try pot for the first time each day, is harder to assess. The number is in the same ballpark as the cigarette smoker number, so on the surface it's not unreasonable, though we might guess that one's first pot experiment might come a bit later, say somewhere around age 16.

I have no personal experience here at all, since marijuana was not even invented when I was a teenager. I've asked many young people for their opinions, but have never gotten a definitive answer. Could the number be right? Do more kids try pot, which is still illegal in most parts of the US, than try cigarettes, which can be bought legally by anyone over age 18 or perhaps 21, and as a practical matter, are readily available to everyone?

One way to help decide is to ask who paid for this advertisement. Again, I don't know where the money came from, but the credit on the page was for the Coalition for a Drug-Free America. Note the difference from the previous set of endorsers: however worthy, the Coalition is a single-issue advocacy group that focuses on trying to reduce drug addiction. It is in their interest to make their issue seem important and worthy of support, and one way to achieve that is to produce impressive and attention-grabbing numbers.

A December 2017 report from the National Institute on Drug Abuse (a US government agency) found that 22.9 percent of high school seniors in their sample had used marijuana in the past month; only 9.7 percent had smoked cigarettes, but 16.6 percent had used some kind of vaping device like e-cigarettes. The overall sample size was 43,700. It's quite possible that the Coalition number is within range.

News media are supposed to be dispassionate neutral reporters of facts, but they can be manipulated, and of course scary headlines attract more readers. As one such headline said,

"UN aid workers raped 60,000 people"
(*The Sun*, February 12, 2018)

The story went on to say that "A whistle blower has claimed UN staff could have carried out 60,000 rapes in the last decade as aid workers indulge in sex abuse unchecked around the world."

As an excellent article by Amanda Taub in the *New York Times* said on March 1, 2018, "That is a horrifying number. That is an attention-grabbing number. It is also more or less a made-up number."

How did this come to pass? A 2017 United Nations report that said it had recorded "311 victims of sexual exploitation by peacekeepers" in the previous year. A former UN employee, now an advocate against such abuse, scaled the number up to 600 to account for both military and civilian personnel, then by a factor of 10 on the theory that only 10 percent of assaults were recorded, then multiplied by 10 again to cover a ten-year period. The *Sun* generated the scare headline from that; it's only by reading much further into the story that the shakiness of the number becomes apparent.

11.2 Why do they care?

"Each year, according to the [American Anorexia and Bulimia] Association, 150,000 American women die of anorexia."

(Naomi Wolf, *The Beauty Myth*, 1990)

This is a striking number; anorexia is clearly a public health crisis. Or is it? Little's Law again comes to the rescue. How many American women die each year? As we estimated earlier, approximately 4 million Americans die annually, half of whom are women. If the quotation is accurate, the 150,000 anorexia deaths would be nearly 10 percent of all women who die.

That's clearly not right. There's no doubt that anorexia and bulimia are serious health problems for many young women, but the original number appears to be a misquotation of information originally provided by the American Anorexia and Bulimia Association, which said that there were about 150,000 *sufferers*; that's quite different from deaths. Whether it was done consciously or not, there was a natural tendency to repeat the large number with the wrong units and the number took on a life of its own, even though a moment's thought would show that it can't possibly be even remotely correct. (Ms Wolf deleted the statement in the paperback version of *The Beauty Myth* published in 1992.)

11.3 What do they want you to believe?

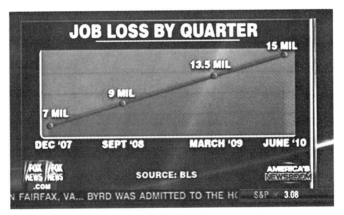

Figure 11.1: Non-uniform horizontal scale.

In Chapter 10, we discussed gee-whiz graphs, where an inflated vertical scale might mislead, and we also mentioned the possibility of something similar on the horizontal axis. Those are less common, at least in my experience, but since they are harder to create, they more strongly suggest an attempt to misrepresent something. The image in Figure 11.1 from a news program is an example of a non-uniform horizontal scale, along with a non-uniform vertical scale and some gee-whiz.

With a bit of work we can redraw this graph to make both axes uniform and set the origin at zero; the result is shown in Figure 11.2.

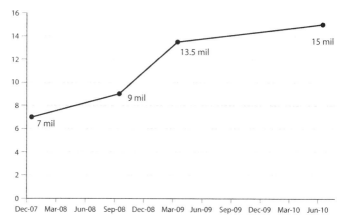

Figure 11.2:   Uniform scales.

Clearly the upward trend is not nearly so smooth as the original suggests, nor are the time periods the same. I don't know whether this graphic was artistic license or an attempt to make a claim about unemployment under a particular president, but it's deceptive either way.

Gun control is another hot-button issue in the US, where powerful interest groups like the National Rifle Association have led to a society where the number of guns in private hands is comparable to the number of people, according to a 2009 survey by the Congressional Research Service.

Deaths by guns are all too common; the number is well over 30,000 per year, and many people are justifiably alarmed:

"The number of American children killed by guns has doubled every year since 1950."

(Nancy Day, *Violence in Schools: Learning in Fear*, 1996)

This remarkable statement, which I first saw in Joel Best's *Damned Lies and Statistics*, doesn't stand up to much scrutiny. Suppose that a single unfortunate child was killed in 1950. Then there would be two in 1951, four in 1952, over a thousand by 1960, over a million by 1970, a billion by 1980, and a trillion by 1990.

Surprisingly, one can still find the story repeated almost verbatim: for instance, *Staying Safe in School, 2nd edition* (Chester and Tammy Quarles, 2011) says "Every year since 1950, the number of American children killed with guns has doubled."

It seems likely that this specific example started originally with a minor slip in transcription; the original may have said "has doubled since 1950" or perhaps the time unit should have been decade instead of year.

One of the oddest graphs that I've seen was published by Reuters in 2014. Figure 11.3 shows the number of firearm murders in Florida over a 20-year period. At first glance it might suggest that the murder rate dropped when the "Stand Your Ground" law was enacted, but it's upside down—the numbers increase downward.

If we plot the data the right way up, it seems to show a rise in murders when the law was implemented, though we certainly don't know whether this is cause and effect or merely correlation.

Gun deaths in Florida

Number of murders committed using firearms

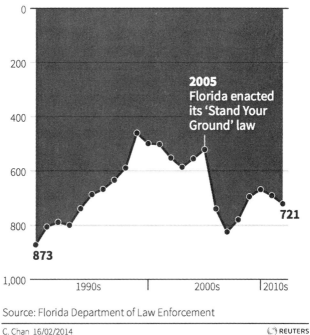

Source: Florida Department of Law Enforcement

C. Chan 16/02/2014   REUTERS

Figure 11.3: "Stand Your Ground" laws and gun deaths.

11.4 Summary

The lesson of this chapter is to think about the source of the information that you've been given, and ask yourself whether the source might have some agenda. This applies to the health risks of smoking, sugar, caffeine, alcohol, marijuana, and pretty much anything else that we might enjoy. It applies to gun control, especially in the US, and to climate change (again, especially in the US). Powerful commercial and governmental interests try to influence us, and have the resources to do an effective job.

We always have to consider the source of information when trying to assess its validity and accuracy. There's a natural tendency to want to emphasize data that makes a desired point while downplaying data that doesn't support a position, and this tendency is even stronger where there's money or politics involved. "Follow the money" is good advice. The more extreme the position, the more likely that some kind of bias is at work—as with miracles and magic, extraordinary claims require extraordinarily good evidence.

Chapter 12

Arithmetic

"I was never any good at math."

(All too many people)

If I had a dollar for every student who has said to me "I was never any good at math," it wouldn't be enough to retire on, but I could certainly treat my family and a few friends to an exceptionally fine dinner somewhere. I think that in most cases, it's not that people can't do arithmetic, but some combination of poor teaching, lack of practice and too little motivation makes them give up without even trying.

It doesn't have to be that way. In this chapter, we'll look at a handful of techniques that will help you do your own arithmetic when it's necessary to assess numbers that someone else has presented, and to produce your own numbers when you need to. There will also be a couple of shortcuts that let you do some kinds of arithmetic more quickly, without resorting to a calculator or even pencil and paper.

As I said earlier, this isn't the "math" that you might have suffered through in school; it's nothing more than basic arithmetic from elementary school, but simpler because you can take shortcuts to make things easier. I think that with a bit of experience, you'll find that freedom liberating.

12.1 Do the math!

> "Toyota is saving $10 a car. That adds up to annual savings of $30.2 million, based on the 302,000 Camry's Toyota sold last year."
> (*New York Times*, January 13, 2006)

Oops! That should be $3.02 million. It's easy to multiply and divide by 2 or 10, and no harder for powers of 10; errors like this one shouldn't occur and should be easy to spot if they do.

You'll get better at arithmetic if you practice. Start with something simple like checking the order of magnitude—the power of 10—when you see implicit computations like this:

> "It's going to cost 200 million dollars a year, and then take between three and four years. So that's almost a trillion dollars!"
> (Bill O'Reilly, *Fox News*, 2010)

Oops! That trillion should have been billion.

> "Cisco's stock exploded, and at the peak of the dot-com bubble in March, 2000, it was the world's most valuable company, with a stock market value of $555 million (U.S.). Some thought it would be the world's first trillion-dollar company."
> (*Toronto Globe and Mail*, December 24, 2017)

Oops! That million should have been billion.

Check percentages when you have the data:

> "Pebble is laying off 25% of its employees—that's 40 pink slips—taking it down to just 80 people."
> (slashdot.org, March 2016)

Oops! If 25% is 40 people, then 100% is 160 people, and Pebble should still have 120 employees left. Something's wrong.

And sometimes there's just no obvious explanation for an erroneous computation:

> "In California, tap water costs around one tenth of a cent per gallon, while bottled water is 0.90 cents a gallon. That makes tap water 560 times less expensive than bottled water."
> (BusinessInsider.com, 2011)

I suspect that "0.90 cents" should have been "90 cents," and thus that tap water is 900 times less expensive than bottled water, but "560" comes out of nowhere.

In all of these, the trick is to get started. Rather than taking some statement at face value, do a quick check of the arithmetic. It only takes a moment to decide either "looks good to me" or "wait a minute...".

12.2 Approximate arithmetic, round numbers

You have undoubtedly noticed that in the process of deciding whether some number is reasonable, we've ruthlessly rounded off to multiples of 2, 5 or 10, and adjusted numbers so they can be easily multiplied or divided by others. We can almost always get away with this because the job is to decide whether some number is in the right ballpark, not whether it's exactly correct. Indeed, for many of the things we've talked about, there's no "exact" or "correct" anyway, since no one knows the values that have gone into a computation with any degree of accuracy.

In a sense, this is the flip side of specious precision. When the numbers that go into a computation are approximate, it makes no sense to produce a precise result. Similarly, if the result can only be an approximation, there's no need for the input data to be highly precise either.

One approach to easy computation is to start with round numbers, adjusting them later if necessary. For example, in different parts of the book, we've used a variety of values for the population of the US, ranging from 300 million to 330 million. No matter what, this is at most a 10 percent difference, so our final conclusions will not be off by more than 10 percent because of an error in this factor. Furthermore, if a round number like 300 million makes arithmetic easier, we can just use it, and then at the end scale the final result by 10 percent. That's simpler than trying to work all the way through with 330 million. Similarly we've assumed life expectancies anywhere from 65 to 80. Again, that's usually good enough, since at most we're off by about 20 percent.

12.3 Annual and lifetime rates

"According to the American Cancer Society, nearly 221,000 new cases of prostate cancer—one out of every six men—will be diagnosed in the US in 2003. An estimated 28,900 men will die from the disease."
    (http://www.endocare.com/pressroom/pc_treatment.php)

"One man in 6 will get prostate cancer during his lifetime. And one man in 35 will die of this disease."
(American Cancer Society)

Notice the difference in these two statements, between the
chance of getting a prostate cancer diagnosis this year versus
the chance of ever getting prostate cancer. The first quote
makes the common error of confusing annual risk with lifetime
risk. It clearly can't be right; there are 150 million men in the
US, so 221,000 is nowhere near one sixth of them. On the
other hand, about 2 million men turn 65 each year. If one in
six gets a positive diagnosis on his 65th birthday, that's about
330,000, which is 50 percent higher than the number given, but
not unreasonable.

It's easy enough to find similar instances for women's
health issues like breast cancer. The following correction
shows an example where the original story confused relative
rates with absolute rates:

"For every 100,000 women in 2010, the mortality rate
for black women was 36, and the mortality rate for
white women was 22, which is about 1.64 black women
for every white woman. It is not the case that nearly 14
black women in Tennessee die from breast cancer for
every white woman who does."
(*New York Times*, December 2013)

Mortality rates are often given as the annual number of
deaths per 1,000 or, as in this case, per 100,000. They represent the expected number of deaths from some disease in a
group of that many people. The statement above says that of
100,000 black women in Tennessee, 36 will die of breast cancer, while for 100,000 white women the corresponding number

is 22. Dividing 36 by 22 gives 1.64, so the risk of dying from breast cancer is 1.64 times higher for black women than for white women in Tennessee, not 14 times higher. I would guess that the number "14" must have come from simply subtracting 22 from 36, rather than computing the ratio.

12.4 Powers of 2 and powers of 10

Many of the numbers that come from technology involve powers of 2, because computers use the binary number system, which uses base 2 instead of base 10.

For the most part, this has little to do with us in daily life, but occasionally the idea surfaces. As it turns out, there's a close relationship between some powers of 2 and some powers of 10, which makes it possible to do certain computations really easily.

If we raise 2 to the 10th power, that is, multiply 2×2×...×2 ten times, the result is 1,024, which we can easily check by continuing the sequence 1, 2, 4, 8, 16, 32, ..., and so on. 1,024 is close to 1,000 or 10^3, about 2½ percent higher. Now look at 2^{20}, which is 2 multiplied by itself 20 times, which is also 1,024×1,024. That's 1,048,576, which is five percent more than a million, or 10^6.

If we do the same thing with 2^{30}, we find it's about 7½ percent over a billion, or 10^9. Every 10th power of 2 is a reasonable approximation to a power of 10 that is going up by 3 each time. The approximation gradually gets worse, but it's good for a surprisingly long distance; for instance, 2^{100} is only 27 percent greater than 10^{30}.

In *The Social Atom* (2007), author Mark Buchanan says "Take a piece of whisper-thin paper, say 0.1 mm thick. Now suppose you fold it in half twenty-five times in a row, doubling its thickness each time. How thick will it be? Almost everyone asked such a question will grossly underestimate the result."

Right now, make your own estimate. You could just multiply 2 by itself 25 times, then multiply by 0.1 mm. But you can use the relationship between the powers of two and powers of ten to simplify the arithmetic: 2^{25} is 2^5 times 2^{20}, and 2^{20} is about a million. This isn't exact, of course, but it's close enough to provide a good estimate.

When you have finished this exercise, you can decide whether you are better than "almost everyone." I'm sure you are, but let's check to be sure.

Brief pause while you work this out for yourself...

The approximation is 32 million times 0.1 mm, or 3.2 million millimeters, or 3.2 km (2 miles). If we had used the exact value of 2^{25}, which is 33,554,432, you can see it wouldn't have made any real difference, especially since the paper thickness of 0.1 mm is only an approximation anyway. That's a useful lesson that I'm going to keep repeating: approximation is your friend—you can get good-enough answers easily, because small errors in one direction may be compensated for by other errors in a different direction.

And with that example under control, here's another. The comedian Steven Wright says, in his usual deadpan style, "I have a map of the United States...actual size. It says, 'Scale: 1 mile = 1 mile.' I spent last summer folding it." Suppose for simplicity that the map is 4,000 km by 4,000 km. How many

times would you have to fold it in half to make it into a square
1 meter on a side? You may ignore the practical reality that
you can't fold a piece of paper more than a few times—this is
only a thought experiment.

If the original map is made of paper 0.1 mm thick approximately how thick will the folded map be?

Is the estimate of paper thickness, 0.1 mm, much too high,
much too low, or about right, and why? You can make your
own estimate of how thick paper is by observing reams of
paper near printers or by thinking about the number of pages in
this book.

12.5 Compounding and the Rule of 72

"In his will, [Benjamin] Franklin left 1,000 pounds ster-
ling to the cities of Philadelphia and Boston, with the
stipulation that the funds be lent out at 5 percent interest
a year. Because of compounding, Franklin figured that
in 100 years his bequests to these cities would be worth
131,000 pounds."

(TIAA/CREF *Participant*, 2003)

Benjamin Franklin died in 1790, so by 1890 his bequests
should have been worth quite a bit, but 131,000 pounds each
seems like a lot. Is it correct?

The *Rule of 72* is a rule of thumb for estimating the effects
of compounding, where some quantity grows by a fixed percentage in each of a series of identical time periods. The Rule
of 72 says that if a quantity is compounding at x percent per
time period, the time it takes to double is approximately $72/x$
periods. For example, if college tuition is rising 8 percent per

year, college will cost twice as much as it does today in 72/8 or 9 years. But if tuition is rising more slowly, say 6 percent per year, doubling will take 72/6 or 12 years. If the inflation rate is 3 percent per year, prices will double in 24 years, or the money you hid under your mattress will only buy half as much 24 years from now.

Conversely, if the doubling time is given, you can compute the rate by dividing 72 by the number of time periods. For example, if the cost of a new car has doubled in the past dozen years, it has been increasing at 72/12 or 6 percent per year. Remember a couple of examples like this and you can always recreate the Rule.

Back to Ben Franklin. At 5 percent per year, the doubling time is 72/5, or about 14 years: every 14 years or so, his bequest would be worth twice as much as it was the previous period. There's room for a little over 7 doublings in 100 years (7 times 14 is 98), and 2^7 is 128, so 1,000 pounds would become 128,000 pounds. Add in a couple more years of accumulation and 131,000 pounds is evidently correct. The precise value, which you could compute with a calculator, is 1,000 times 1.05^{100}, or 131,501.

When money compounds, it grows more rapidly than just in proportion to time, because the interest in each period is added to the total being invested for the next period, a fact that is not always appreciated. I once heard a story on National Public Radio that said that getting 20 percent interest for five years doubles your money. That's only true if you put the interest under your mattress each year for five years so it doesn't compound. The Rule of 72 says that the doubling time at 20 percent is about 3.6 years; the true value is a bit longer (3.8 years). If you do compound 20 percent for five years, the value will be almost 2.5 times the original, better than NPR's figure.

One must be careful of the difference between compounded change and linear change. For example,

> "The Alpine glaciers are losing one percent of their mass every year and, even supposing no acceleration in that rate, will have all but disappeared by the end of the century."
>
> (Climate change web site, 2010)

The Rule of 72 tells us that if the glaciers shrink by one percent a year, they will be only half gone in 72 years, and only three quarters gone in 144 years. Of course this is a gross oversimplification of the physical processes, so the original claim may well be true; it's just that we can't conclude that from incorrect arithmetic.

And while we're thinking about long-term compounding,

> "Thus, every £1 which Drake brought home in 1580 [accumulating at 3½ percent compound interest] has now become £100,000. Such is the power of compound interest!"
>
> (John Maynard Keynes, "Economic Possibilities for our Grandchildren," 1928)

You might apply the Rule of 72 to judge the accuracy of Keynes's computation.

The Rule of 72 approximation breaks down if the rate is too high, but it's fine for the sorts of rates and time periods that we encounter in daily life. It also assumes that the compounding is uniform across the whole time. Uniformity is often a good assumption, or at least good enough to come up with reasonable answers.

12.6 It's growing exponentially!

> "The number of people on the Internet on average has
> doubled every year for the last 11 years and is expected
> to keep growing exponentially for the next decade or
> more."
>
>   (Environmental web site, 2001)

When this statement was written in 2001, there might have
been 100 million people using the Internet. If that number
continued to double every year until 2011, there would be over
100 billion people using the Internet. Since that exceeds the
population of the earth by a factor of well over ten, it's
unlikely.

Even if my estimate of "100 million" in 2001 should have
been "10 million," a decade of annual doublings would result
in 10 billion users, and we're not there yet either. The growth
could still be exponential, but the doubling time would have to
be longer than "every year."

There are two useful lessons here. One is terminology: the
word "exponentially" has come to mean "growing fast," and
precise quantitative meaning has disappeared.

> "The energy capacity of batteries is increasing 5 percent
> to 8 percent annually, but demand is increasing expo-
> nentially."
>
>   (Newspaper story on batteries, 2006)

"Exponential" growth means compounding, plain and simple. If capacity is growing at 8 percent annually, that's an
exponential growth, and capacity will double in about 9 years.
At 5 percent, the doubling time is more like 14 years, but the
growth rate is still exponential.

The other lesson is that any true exponential can't go on for-ever: something runs out.

> "The past 30 years all we have been doing is ramping up anti-drug efforts across the country. The budget for the war on drugs has doubled every year since its inception during the Nixon administration."
>   (web site, circa 2005)

The Nixon administration ended in 1974, so let's call it 30 doublings up to 2005 when this was posted. Recall that 2^{30} is about a billion. Even if Nixon's original budget was merely a token like $1,000, we're talking about over a trillion dollars today.

Estimates of the actual cost of the war on drugs vary, naturally, but there's a consensus around $30 billion per year to fight drugs. Perhaps the author meant "every decade"? That would only be a factor of eight, which seems small but certainly possible.

> "The number of centenarians in Switzerland has doubled every year since the 1960s."
>   (*Iran Daily*, 2015)

Even if there were only one Swiss centenarian in 1965, there would be 2^{50} of them in 2015; that's a biggish number for a small country. The story goes on to say that "In 1941, the tally of people aged 100 or more was 17; in 2001, it was 796." That is, in 60 years, the number went up by a factor of 47. If the time interval should have been decades instead of years, that would imply six doublings, or a factor of 64, which is within range here.

12.7 Percentages and percentage points

> "They say it will save $10 million of their $7.8 billion budget. That's a wee bit more than a 1/1000th of 1 percent."
>
> (*Newark Star-Ledger*, January 7, 2015)

It's easy to make a slip when working with percentages, because there's a factor of 100 floating around; apply it wrong, and you're off by a factor of 100. In the line above, let's round 7.8 up to 10 to make the arithmetic easier. Say the budget is $10 billion. One percent of that is (literally) one one-hundredth, or $100 million. One one-thousandth of that is $100,000, not $10 million. It looks suspiciously like the author of this quotation meant to say "1/10th of 1 percent."

It's easy to misplace a factor of 100, but sometimes a quick check will help you to find it again. For example,

> "out of the approximately forty-five hundred recipes that appear in later editions, he'd chosen eighteen, a mere 0.004 per cent of the book's content."
>
> (*New Yorker*, March 21, 2018)

Always go for the easy arithmetic first. One percent of 4,500 is 45, so 18 must be somewhat less than half a percent, or 0.4 percent. The original value of 0.004 is wrong by a factor of 100.

A *percentage point* is a difference of 1 between two percentages, for example between 5 percent and 6 percent. Percentage points provide another example of a place where language can trip us up, similar to the distinction between 1 degree Celsius and 1 Celsius degree. An article in the *Los Angeles Times* said in December 2010 that President Obama's tax cut package would reduce the Social Security payroll tax by 2 percent; this was later corrected to say "a drop of 2 percentage points, reducing withholding from 6.2% to 4.2%," which drops the withholding by about 1/3, or 33 percent.

An article in the *New York Times* in September 2006 said that a state sales tax rose from 4 percent to 6 percent, and called it an increase of 2 percent. This is really an increase of 2 percentage points, not 2 percent, so it's a 50 percent increase in the sales tax. Percentage points are confusing; watch out when others use them, and avoid using them yourself.

Shifting between fractions and percentages can also lead to confusion:

> "The stock market is falling out of favor as a way to generate income in retirement. Only 1 in 5 current workers think stocks and stock mutual funds will provide a significant amount of their retirement income, down from 24 percent in 2007."
>
> (Financial advice web site)

"1 in 5" is 20 percent, which is a modest drop from 24 percent, but not exactly "falling out of favor." It would be clearer to say that 24 percent of workers used to prefer investing their retirement funds in stocks, and now it's 20 percent.

12.8 What goes up comes down, but differently

> "The Harvard endowment organization grew by 33% under the previous manager, but the new manager is going to cut 25%."
>
> (*New York Times*, February 7, 2009)

So Harvard's endowment organization (which presumably manages Harvard's many-billion-dollar endowment) is going to wind up with a net growth of 8 percent. Or is it?

This is a nice example of a problem frequently encountered when talking about percentage changes that go up and down: if what goes up does come down, it comes down by a different percentage.

To see this, let's start with a specific number, which is usually a good way to begin. Suppose that the endowment organization began with 75 people, a number that will make our subsequent arithmetic easy. Growing 75 by 33 percent adds 25 people, which takes the staff to 100. When the new manager arrives, 25 percent of the 100 people get fired—that is, 25 people—and Harvard is right back where it started.

This somewhat counter-intuitive result happens because the second percentage is based on the new value, not the original one. To pick an example from another domain, if the price of some stock goes down by 50 percent, which is not unheard of, it has to rise again by 100 percent to get back to its original value. Investors do not always understand this unfortunate fact.

Consider the graph of total returns in Figure 12.1, which was published by a mutual fund some years ago. Each bar represents the percent gain or loss for a particular year compared to the previous year.

Suppose that we start with an investment of $1,000 at the beginning of 2000. Then at the end of the first year our investment is worth $1,075, and at the end of the second year it's worth $1,021 (that is, 95 percent of 1,075).

Following along for each year, at the end of 2007, our investment is worth $1,476, which is pretty good.

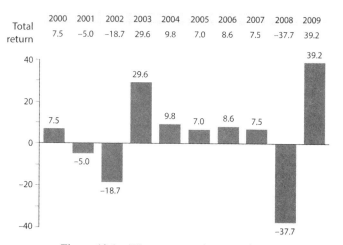

Total return	2000	2001	2002	2003	2004	2005	2006	2007	2008	2009
	7.5	−5.0	−18.7	29.6	9.8	7.0	8.6	7.5	−37.7	39.2

Figure 12.1: What goes up often goes down.

Unfortunately 2008 was a terrible year for all investors, and the 37.7 percent decline takes our portfolio down to $919, below what we started with nine years earlier!

The 39.2 percent rebound in 2009 brings us back up only to $1,280, about where we were at the end of 2005. (All of this ignores any effects from inflation.) What goes down by a particular percentage does not recover when it then goes up by the same percentage. Be careful about the base for a percentage computation.

> "According to 2008 Census data, the median earnings of someone with a bachelor's degree was $47,853, or 43 percent higher than the $27,448 earned by someone with only a high school diploma."
>
> (Advertisement for Excelsior College, 2010)

The ratio of 47,853 to 27,488 is 1.74, so the bachelor's degree earnings are 74 percent higher, not 43 percent. On the other hand, the ratio of 27,488 to 47,853 is 0.574, so high school diploma holders only earn 57 percent of what BAs do. The "43 percent" in the original presumably comes from subtracting 57 from 100.

The proper conclusion is that people with only a high school diploma earn about 43 percent less than those with a bachelor's degree.

12.9 Summary

Most of the arithmetic one needs in day-to-day numeric self defense is straightforward, not much more than multiplication and division. You can get better at this with practice. For example, how do you compute a tip in a restaurant? Let's say that the bill is $50. Don't use an app on your phone. Instead, compute 10 percent by moving the decimal point over; that's $5. For a 20% tip, double that to $10. For a 15% tip, add half of it, for $7.50. For an 18% tip, drop the 20% tip by ten percent to $9. Round up or down as you see fit.

Approximation is your friend. You can safely round values to nearby numbers that are easier to work with. A sequence of approximations often converges to a good answer. It's not always the case, but if one approximation is too high there will frequently be some compensating approximation that's too low to help balance it out.

You can always be conservative, rounding up to make sure an estimate is large enough, or down to ensure it's small enough. For instance, we have approximated US life expectancy as 75 years. If this estimate is too low, then the number of deaths per year we compute will be too high (because we should be dividing by say 80 instead of 75). The actual life expectancy in the US is nearly 79, so our estimates of deaths, people turning 65, and so on have been too high by about 5 percent (79/75).

There are a handful of numeric rules of thumb that help, notably the Rule of 72 when compounding is involved.

Watch out for percentages; it's easy to misplace a factor of 100 if you're not careful about whether a number expresses a percent or a fraction, and you have to pay attention to the base that you're using to compute a percentage.

Chapter 13

Estimation

"Fifty billion plastic water bottles are discarded annually
by Americans; 20 billion barrels of oil are used to make
this plastic and 25 million tons of greenhouse gases
are released into the atmosphere."
(Environmental site blog post, September, 2015)

We've spent a fair amount of time assessing numbers that other people have given us, often finding significant errors along the way. (Of course this is sample bias at work; the majority of instances where the numbers are accurate aren't as interesting or instructive for this book.)

We have not spent as much time coming up with numbers of our own, working from common sense and our own experience. Let's do a few of those, starting by trying to create an independent estimate of the number of plastic water bottles used in the US each year. Making your own estimate first is good practice and is often a great way to begin assessing someone else's numbers, like the ones above.

Go ahead—do your own estimate, using what you know from your own life.

13.1 Make your own estimate first

This seems like a case where working from the bottom up is best. How many plastic water bottles do you use in a typical week? My own use is limited since I'm rarely in an environment where I have to carry water around, the local water supply is fine, and there's a water cooler and filter down the hall from my office. I haven't counted, but I would guess that my use, averaged over a year, is around one bottle a week.

Think about your own usage and how it compares to people you know. What's a typical value in your experience? A reasonable range might be somewhere between one a day and one a week, with plenty of outliers. At one bottle per week, that's 50 per year per person, and thus about 15 billion per year for the whole country. One per day is about 100 billion per year.

Thus a defensible estimate would be somewhere between 15 and 100 billion. We could just average these values, which would be about 60 billion, but in practice, it's often better to use the *geometric mean*, which is the square root of the product. In this case, that's the square root of 1,500 billion billion, or about 40 billion. The geometric mean is better because with the arithmetic mean the much larger value dominates. Think about the average of a thousand and a million: it's half a million. The geometric mean is 30,000. If we're uncertain about the values on both ends, the geometric mean is better.

With this ballpark estimate in hand, "fifty billion" sounds reasonable, especially given uncertainty about whether "discarded" might mean "discarded or recycled." It's also consistent with stories that say that Americans drink 9 billion gallons of bottled water each year, since a gallon would be around 5 or 10 bottles.

While we're on the topic, how about the "20 billion barrels of oil" needed to make the bottles? If it takes 20 billion barrels of oil to make 50 billion bottles, that's 4/10 of a barrel to make one bottle. As we saw in Chapter 2, an oil barrel is 42 gallons so this implies that it takes nearly 17 gallons of oil to make a single plastic water bottle! Even if this number includes all the manufacturing and shipping costs for bottled water, it must be much too high. Could it be our old friends millions and billions again?

We can do the arithmetic (which I won't bore you with), assuming 20 million barrels, to determine that it takes about two ounces of oil to make a plastic bottle that weighs about one ounce. I don't know enough about the manufacturing process to assess this myself, but it's consistent with data found on a variety of web sites.

You might also think back to one of the examples in Chapter 1, where we figured out that US oil consumption for cars was 2.5 or 3 billion barrels annually. Does it seem likely that we would be using 6 or 7 times as much oil to make plastic bottles as we do to drive around?

How about the third number in the original quote, "25 million tons" of greenhouse gases that are released into the atmosphere during manufacturing? Doing the arithmetic says that producing one plastic bottle produces one pound of greenhouse gases. I honestly don't know about that—though it seems high, the frequent error of mixing up tons and pounds would lead to a value that seems too low. To resolve this one, we need more information.

13.2 Practice, practice, practice

The best way to get better at estimation is to practice. Everyday life provides an endless variety of opportunities, and if you try a few regularly, you'll get better fast. It's also kind of fun in a nerdy way.

Here's an example that I'm fond of. The picture in Figure 13.1 shows a cannon that sits in front of a building at Princeton University. The story is that it was left behind by George Washington after the battle of Princeton in 1777, relocated 15 miles north to New Brunswick during the War of 1812, then repatriated to Princeton in 1838. Most students pass by it several times a week, even several times a day.

They see it, and yet they don't. For years now, I have asked students in my classes to estimate how much the cannon weighs. Since you probably haven't seen it in the flesh, here are a few facts. It's about 10 feet long, 24 inches in diameter at the fat end, and 15 inches at the muzzle. It fires what was probably a 6-inch ball.

How much does it weigh? Take a moment to come up with your estimate, and then we'll talk about it.

I've asked students this question for years, and have collected a wide range of answers. The largest estimate so far is 300,000 pounds (!!) and the smallest is 50 pounds (!!!). What's a reasonable value?

My personal estimate says the cannon is 10 feet long, and on average about 1 foot by 1 foot if I ignore the hole in the middle, so it's about 10 cubic feet of whatever it's made of. In the 1700s, that was cast iron. One of the useful engineering facts that I still recall from college is that the density of iron is about 450 pounds per cubic foot, so the cannon would weigh about 4,500 pounds.

Figure 13.1: The cannon in front of Cannon Club, Princeton, NJ.

If you prefer metric units, it's very roughly 3m by 1/3m by 1/3m, or 1/3 of a cubic meter. The density of cast iron is about 7,500 kg per cubic meter, so the weight is about 2,500 kg, or about 5,500 pounds. The two estimates are within about 20 percent—close enough, given how casual I was about the dimensions.

What if you don't have a clue about what the cannon is made of, let alone its density? Well, you can be sure that it's denser than water, or cannons would float. It's probably quite dense, or it wouldn't require soldiers and horses to move cannons around. The density of water is a little over 60 pounds per cubic foot, a useful number to know. So if cast iron is five times as dense as water, that would imply a weight around 3,000 pounds.

The extreme values that I quoted above are likely from students who just wrote down a number because one was required, not the result of any actual thought. Reasoning backwards: if a cannon weighed only 50 pounds, an ordinary soldier in the Revolutionary War could have carried one under his arm, at least for a while.

You've undoubtedly heard the phrase "the wisdom of crowds," the idea that if a group of people make independent estimates of something, their average estimate will be pretty accurate.

That's been my experience with the cannon: although the outliers are sometimes very far off, the median has been around 2,000 pounds—too low but not ridiculously so—and the average more like 5,000 pounds—more accurate but in part because of heavy outliers. Figure 13.2 shows the estimates produced by one of my classes, sorted into increasing order by weight. The median is 2,000 pounds and the average is 4,240.

I genuinely don't know the actual weight. I once asked a friend in the History department, who said he didn't know either; as he told me, "Historians don't measure, we tell stories," though I think that for rhetorical effect he was downplaying how important quantitative data is to historians. A friend who is interested in military history did some digging and as far as he and I can see, the gun is a British 24-pounder and likely weighs about 5,000 pounds.

13.3 Fermi problems

> "How many piano tuners are there in Chicago?"
>   (Attributed to Enrico Fermi)

Enrico Fermi was an Italian-born physicist who emigrated to the US in 1938 to escape fascism. He won the 1938 Nobel

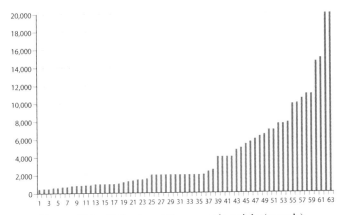

Figure 13.2: Estimates of the cannon's weight (pounds).

prize in physics, built the first nuclear reactor in 1942 at the University of Chicago, and was a crucial member of the Manhattan Project team that created the first atomic bomb in 1945.

One of Fermi's many talents was his ability to make remarkably good estimates of quantities for which there simply wasn't enough information. Today, such estimation problems are called *Fermi problems*, and estimating the number of piano tuners in Chicago is the archetypal example. They are also sometimes called "back of the envelope" problems, since you should be able to solve them with nothing more than a pencil and a small piece of paper.

Fermi problems are frequently used in physics and engineering courses to teach students how to make justifiable assumptions and sensible approximations, while keeping dimensions correct. Most such problems are more specialized and technical than we meet in ordinary life, but the spirit and approach is the same. The big difference is that we don't need to make nearly as many educated guesses about unknowns before getting to an answer.

Here are a handful of examples that I've used over the years, or picked up from others. Students tell me that questions like these sometimes occur in job interviews, especially for quasi-technical positions like finance or consulting, so practice has been helpful. As you read them, come up with your own estimates; mine will follow.

• How many people could fit in a given space, for example a football or soccer field, if they are standing at normal distances from each other? This is good practice for figuring out crowd sizes at public events like rallies and protests, where official estimates are sometimes disputed. (Donald Trump's estimates of the crowd for his inauguration in January 2017 were two or three times bigger than those from more dispassionate sources.)

• How many leaves do I rake each fall when I have to get them off my lawn? It always feels like billions, but I can pass some of the time by making an estimate while I work. I have half a dozen big oaks and maples.

• How many petabytes would fit in the room where you are right now, if they were stored on standard laptop disks like the one in Figure 13.3? Ignore wires, power, and the like. I use this question in a course about basic computing for non-technical students.

• What's the surface area of your body?

• How much cash does an armored truck hold? I borrowed this from a draft book on estimation by Sanjoy Mahajan at Olin College of Engineering.

Figure 13.3: Some gigabytes for a laptop.

• How many golf balls can you fit into a school bus? Purportedly a Google interview question, though every Googler I've asked doubts it, at least for technical positions.

• How many miles did Google drive in your country to take all the pictures for Street View? How much gas was used? How long did it take? How much data was stored? How much did it cost?

13.4 My estimates

Have you tried these yourself first? It's good practice, and it's a sanity check—if our answers are wildly different, something is awry, and it would be good to understand why. As you read along, notice how casual the arithmetic is, and ask whether there would be a big change if it were done more precisely.

• How many people? If people stand a yard or a meter apart, then each occupies a square yard or a square meter. A US football field that's 100 yards long by 50 yards wide would hold 5,000 people. Tighter packing would increase the number; from our discussion of area in Chapter 6, you should be able to easily compute what the change would be.

• How many leaves? Suppose a tree is a big box about 40 by 40 by 40 feet, with the top and sides covered with leaves. (Leaves need sunlight, so they're on the outside, not the interior.) The surface area is 5 times 40 times 40, or 8,000 square feet. If a leaf is 4 inches by 4 inches, 10 of them would cover a square foot, so each tree has 100,000 leaves. With half a dozen trees, I probably rake well under a million leaves, but it sure feels like far more.

Figure 13.4: A billion leaves?

• How many petabytes? I'm sitting in a room that's about 15 by 15 by 8 feet, so call it 2,000 cubic feet. A disk is about 3 inches by 4 inches, so 10 of them would cover a square foot. If each is 1/4 inch high, that's 50 to the vertical foot, so 500 in a cubic foot, times 2,000 cubic feet, gives a total of 1,000,000 disks (10^6). If a laptop disk is one terabyte (10^{12}), then the room holds 10^{18} bytes: 1,000 petabytes, or 1 exabyte. If the disks have less capacity, the result will be smaller; for instance, with 500 GB disks, there will be half an exabyte.

• Body surface area? For purposes of this estimate, I think of myself as a rectangular solid 2 meters tall and 1/4 meter wide; ignoring top and bottom, that's four surfaces each of which is half a square meter, so I'm 2 square meters in total. This is obviously a gross over-simplification, but I give you my word that I did this computation exactly as you see it, then as a check went to Google, where the first hit (medicinenet.com) says "Average body surface area for adult men: 1.9 m². Average body surface area for adult women: 1.6 m²." You can experiment with other body shapes to see how the estimates change with more or less detail and fidelity.

• How much money? This is similar to the petabyte question. A stack of 50 bills is about 1/4 inch thick and there are about 12 stacks per square foot, so there would be 2,000 bills in a layer an inch thick, or 20,000 bills in a cubic foot. If an armored truck is say 5 by 5 by 10 cubic feet, it holds 5,000,000 bills. If they are US 20-dollar bills, that's $100 million. Of course this analysis ignores weight; this much cash might weigh too much for the truck to hold, and a getaway car would have even less capacity. Mahajan gets a similar value, and in addition quotes some helpful confirmatory data, like the fact that a typical armored-truck robbery makes off with $1 million to $3 million.

• How many golf balls? An American school bus is very roughly 30 feet long and 6 by 6 inside, so call it 1,000 cubic feet. A golf ball is a cube about one inch on a side (this is an approximation, after all), so about 2,000 of them would fit into a cubic foot; the total for the bus is thus about 2 million. It's fun to ask this question of young children, whose experience of school buses is much more recent than mine. They get slowed down by digressions like "Have the seats been removed?" but do well once they get into the spirit of things.

• How much did Google drive? As a crude estimate, the US is 3,000 miles wide and 1,500 miles high. If there's a road every mile in each direction, that's 1,500 east-west roads that are 3,000 miles long and 3,000 north-south roads that are 1,500 miles long, for a total of 9 million miles. This simplistic model is obviously way too sparse for cities but perhaps not too bad for substantial parts of the middle of the country. (Informal conversations with Google friends suggest that this estimate is too high but not by more than a factor of three.) You can do your own computation of the other quantities, based on the price of gas, experience with digital cameras or on your phone, and so on.

13.5 Know some facts

Your estimates will be better if you can base them on actual knowledge; for that reason, it's valuable to have a variety of physical constants and conversion factors in your head—how much things weigh, how big are they, how long they take.

My working list is summarized in Figure 13.5. Besides the values listed there, I remember a hodge-podge of factoids like populations and areas of various geographical regions, and a bunch of random dates. Your list will undoubtedly be different, though fundamental weights, measures and conversion factors are likely to be important to everyone. As you do more estimation, you'll build up a collection and that will help you get better.

1 gallon of water weighs 8 pounds
1 cubic foot of water weighs 60 pounds
1 cubic foot of rock or concrete is 200 pounds;
    loose dirt is 100 pounds; metals are 400 pounds
1 liter is just over 1 US quart
1 kilogram is 2.2 pounds
1 ton is 2,000 pounds; 1 metric ton is 1,000 kg or 2,200 pounds
1 meter is a little over 3 feet or 1 yard
1 centimeter is 4/10 of an inch
1 mile is 1.6 km
MP3 music is 1 megabyte per min.; CD audio is 10 MB/min.
Electricity costs 10-20 cents per kilowatt hour
Speed of light is 1 foot per nanosecond
Speed of sound is 1,000 feet per second
60 miles per hour is 88 feet per second
100,000 seconds in a day, 30 million seconds in a year
250 working days and 2,000 working hours in a year

Figure 13.5: Some useful approximate numbers.

13.6 Summary

Estimation is easier than you might think before you try it. It's easy because the arithmetic is very approximate, errors tend to cancel each other out, and your assumptions don't have to be at all exact, just vaguely sensible.

After you make an estimate, try to check it either by making another estimate with independent assumptions and computations, or by checking with online sources. But doing it yourself first is the best way to get better. Once you get into the habit, you'll improve rapidly, and you might even find that it's a rewarding game.

One of my friends routinely keeps a running mental total of his groceries, to the nearest dollar or two, as he puts items into his shopping cart. At checkout, if his total differs too much from the cashier's, something might have been charged twice or missed entirely. Sometimes this saves him money but even if not, it flexes his arithmetic muscles—it's good practice. And it's a good example of how approximate arithmetic works: rounding prices up or down to the nearest dollar averages out, so the overall error won't likely be more than a dollar or two.

Chapter 14

Self Defense

"Innumeracy, an inability to deal comfortably with the
fundamental notions of number and chance, plagues
far too many otherwise knowledgeable citizens."
(John Allen Paulos, *Innumeracy*, 1988)

We've come a long way in the past 13 chapters, and I hope
that you feel better able to deal comfortably with fundamental
notions of number and chance. Now it's time for a brief wrap-
up and some general advice before you head off, well armed to
defend yourself.

14.1 Recognize the enemy

Watch for the warning signs that indicate some number or
computation or conclusion is suspect, worthy of skeptical
attention.

If my sampling is anything to go by, there are millions,
maybe even billions, of errors that confuse millions and

billions and all the other powers of 1,000. When you see a big number that seems too high or too low, try bringing it down to its effect on you personally: estimate your share of the big number and relate it to your life and experience. That often makes it easier to assess plausibility. If your share of the national debt or the budget could be taken care of with the small bills in your wallet, something's wrong.

Excessive precision is another suspicious sign. In day-to-day life, it's somewhere between hard and impossible to get precise values for many quantities—incomes, revenues, costs, budgets, rates of change, populations—and thus any such value that is presented with a large number of ostensibly significant digits is surely not as precise as its purveyor would like you to believe. The excess precision may be an attempt to impress, or it may come from blind use of a calculator, and is often related, at least in the US, to mechanically converting from metric to English units. After a while, you'll recognize common conversion factors and know exactly what's going on.

Watch out for arithmetic errors. It's all too easy to slip up when doing computations: if you use a calculator or your phone, a single fat-finger mistake can render an entire computation meaningless. If you already have an order of magnitude estimate for something, however, that acts as an independent sanity check on your arithmetic, so before you start computing, think about what the answer should be. You should be able to estimate at least to within a factor of 10.

Of course wrong units and wrong dimensions go along with this. There's a big difference between a foot and a mile, or a gallon and a barrel, or a day and a year. We've seen many instances of such errors along the way. Sometimes you can detect them by reasoning backwards: if the erroneous unit is sufficiently different from the correct one, it will lead to nonsense results.

Similar advice applies to dimensionality errors: watch out for confusion between square something and something square. That's the most common and easy case, but you can spot others by thinking about the dimensions involved. Area is the product of two lengths, so it has units of length squared, and volume is length cubed. You can often detect such problems by ignoring the numbers and simply checking that the units are correct.

14.2 Beware of the source

Although many of the numeric problems we've seen are merely the result of careless errors or lack of thought, a few have definitely been intended to mislead or misrepresent. So it is always wise to think about the source of your information. What axes are they grinding? What are their motives? What do they want you to believe? What are they trying to sell you? Who paid to send you the message?

Misrepresentation could take the form of bad statistics, or deceptive presentations like some of the graphs we saw in Chapters 10 and 11. It definitely can include statistical flaws like sample bias or survivor bias. Thinking that correlation implies causation has led many people astray quite innocently, and is certainly a core technique for consciously misleading or getting people to believe something that isn't proven and perhaps isn't even true.

Thus the vital questions are: Who says so? Who are they? Why do they care? Where did their data come from? How did they get to their conclusions?

When you do see the data, ask yourself how they know. How *could* they know? Many things can't be known with any certainty, and some can't really be known at all, so you should be wary of people who claim accuracy and precision when rough approximation is the very best that could be achieved.

It's hard for laymen to assess complex technical questions like climate change or the health effects of various substances, but wariness about sources is helpful. The Latin phrase *cui bono* (who benefits?) is just as useful today as it was when Cicero used it over 2,000 years ago.

14.3 Learn some numbers, facts, and shortcuts

You will be much better at checking "facts" from other people if you know real facts yourself. At the very least, it's a help to know some populations, rates, sizes, areas, and so on. I've seen plenty of lists of "numbers you should know," and I have my own steadily evolving collection, many of which have appeared in earlier chapters.

It's helpful to know the approximate population of the earth (call it 7 or 8 billion, depending on which way you want to round) and of your own country, state or province, and town or city. Avoid parochialism by knowing similar values for other countries and cities as well. I've found it useful to have an idea of the areas of various countries and cities as well.

Physical constants and conversion factors are always good to know. Living in the US, it's mandatory to be able to convert between English and metric units, though as we've seen, doing that blindly leads to excessively precise numbers, and sometimes outright errors.

Learn to do approximate arithmetic, which lets you check other people's work quickly. I once came across this sentence: "In this book, $2 \times 2 \times 2$ is almost always 10." It's a good way to think about simplifying arithmetic: an error of 25 percent won't make much difference, especially if it might well be canceled out by another similar error in the opposite direction. In a similar vein, a friend told me of two simplifying equations he learned in physics: 2 equals 1, but 10 does not equal 1.

Among the arithmetic tricks and shortcuts, remember Little's Law, the Rule of 72, and the powers of 2 for repeated compounding.

Get comfortable with scientific notation. It's the best way to deal with big numbers, much better than trying to cope with wordy compound phrases like "million million trillion." In a technological world, it's good to know prefixes like mega, giga and tera as well.

14.4 Use your common sense and experience

In the final analysis, your best defense is your own brain. Common sense will carry you a long, long way in defending yourself, especially if augmented by facts about the real world and your own experience and intuition.

Ask yourself: is that number much too big, or much too small, or could it be about right? Does it make sense? If it were true, what would it imply?

Make your own estimates. No matter how rough, they will guide you in assessing what others say, you'll get better with practice, and you'll have some fun doing so.

Further Reading

There are some fine books about (in)numeracy. My all-time favorite is *How to Lie with Statistics,* by Darrell Huff. It was published in 1954, and is still well worth reading today. If you read no other book on the topic, this is the one.

Joel Best, a sociologist at the University of Delaware, has written three good books on the topic: *Damned Lies and Statistics* (2001), *More Damned Lies and Statistics* (2004), and *Stat-Spotting* (2008). The sub-titles ("Untangling numbers from the media, politicians, and activists," "How numbers confuse public issues," and "A field guide to identifying dubious data") tell where the author is coming from. Best was my original source for the children and guns example in Chapter 11, which has since been described in a variety of other places.

Charles Seife's *Proofiness* (2010) is first rate. The title is a play on "truthiness," a word coined by the American satirical TV program *The Colbert Report*. Says Wikipedia, "*truthiness* refers to the quality of preferring concepts or facts one wishes or believes to be true, rather than concepts or facts known to be true." Proofiness is the same thing, with numbers.

John Allen Paulos's *Innumeracy—Mathematical Illiteracy and its Consequences* was published in 1988, and remains an excellent resource today. Paulos did not coin the word "innumeracy" (which dates from 1959 if not earlier) but his book made it part of our lexicon and helped to sensitize people to the costs and risks of not understanding basic arithmetic and statistics. I also like his *A Mathematician Reads the Newspaper* (1996).

Guesstimation—Solving the World's Problems on the Back of a Cocktail Napkin, by Lawrence Weinstein and John Adam (2008), has a large number of interesting estimation problems, each described on a single page with a solution on the next page. If you like Fermi problems, you'll enjoy *Guesstimation*. A second edition, *Guesstimation 2.0*, was published in 2012.

What If?: Serious Scientific Answers to Absurd Hypothetical Questions (2014), by Randall Munroe, author of the online cartoon *xkcd*, is great fun, with wonderful examples of how to make sensible estimates about some really strange questions. ("How many Lego bricks would it take to build a bridge capable of carrying traffic from London to New York?")

Visit the book's web site at millionsbillionszillions.com for more examples and advice.

Figure Credits

1.1. Courtesy of Minesweeper, CC by SA 3.0.
2.1. Drawing by Emma Burns.
3.1. Drawing by Emma Burns.
4.1. © Ad Meskens / Wikimedia Commons.
5.1. Photo by Waqas Usman.
6.1. Source: Brian W. Kernighan.
6.2. Source: Brian W. Kernighan.
6.3. Source: Brian W. Kernighan.
6.4. Drawing by Emma Burns.
7.1. Source: Brian W. Kernighan.
7.2. Source: Brian W. Kernighan.
8.1. Image by Meghan Kanabay.
8.2. TL, Le Mont-Blanc depuis le village de Cordon, 10/2004. http://artli-bre.org/licence/lal/en/.
8.3. Image by Meghan Kanabay.
8.4. Source: Brian W. Kernighan.
8.5. Source: Brian W. Kernighan.
8.6. Rhymes With Orange © Hilary B. Price – Distributed by King Features Syndicate, Inc.
8.7. DILBERT © 2008 Scott Adams. Used By permission of ANDREWS MCMEEL SYNDICATION. All rights reserved.
9.1. Image Credit: Mark Zuckerberg/Facebook.

9.2. Randall Munroe, xkcd. This work is licensed under a Creative Commons Attribution-NonCommercial 2.5 License. Source: http://xkcd.com/522/.

10.1. Source: Brian W. Kernighan.

10.2. Source: Brian W. Kernighan.

10.3. Data from SEC S-1, October 2013.

10.4. Data from SEC S-1, October 2013.

10.5. Data from National Center for Health Statistics.

10.6. Source: Brian W. Kernighan.

10.7. Data from Fox News.

10.8. Source: Princeton University press release, 2016.

10.9. Source: Brian W. Kernighan.

10.10. Source: Brian W. Kernighan.

10.11. *Graduate News*. Summer 2001 issue.

10.12. Ebirim, C., Amadi, A., Abanobi, O. and Iloh, G. (2014) "The Prevalence of Cigarette Smoking and Knowledge of Its Health Implications among Adolescents in Owerri, South-Eastern Nigeria." *Health*, **6**, 1532–1538. Copyright © 2014 Chikere Ifeanyi Casmir Ebirim, Agwu Nkwa Amadi, Okwuoma Chi Abanobi, Gabriel Uche Pascal Iloh et al. This is an open access article distributed under the Creative Commons Attribution License, which permits unrestricted use, distribution, and reproduction in any medium, provided the original work is properly cited.

10.13. Source: American Cancer Society, Inc. Surveillance Research – 2012.

11.1. Data from Bureau of Labor Statistics.

11.2. Source: Brian W. Kernighan.

12.1. Data from American Funds.

11.3. Source: Reuters/Florida Department of Law Enforcement http://graphics.thomsonreuters.com/14/02/US-FLORIDA0214.gif.

13.1. Photo by Dimitri Karetnikov.

13.2. Source: Brian W. Kernighan.

13.3. Source: Brian W. Kernighan.

13.4. Source: photoeverywhere.co.uk, CCA 2.5 license.